DESIGNING THE TOTAL AREA NETWORK

About the Wiley — BT Series

The titles in the Wiley-BT Series are designed to provide clear, practical analysis of voice, image and data transmission technologies and systems, for telecommunications engineers working in the industry. New and forthcoming works in the series also cover the Internet software systems, solutions, engineering and design.

DESIGNING THE TOTAL AREA NETWORK

INTRANETS, VPN'S AND ENTERPRISE NETWORKS EXPLAINED

Mark Norris
Norwest Communications, UK

Steve Pretty
BT, UK

JOHN WILEY & SONS, LTD
Chichester • New York • Weinheim • Brisbane • Singapore • Toronto

Other Wiley Editorial Offices

John Wiley & Sons, Inc., 605 Third Avenue,
New York, NY 10158-0012, USA

WILEY-VCH Verlag GmbH
Pappelallee 3, D-69469 Weinheim, Germany

Jacaranda Wiley Ltd, 33 Park Road, Milton,
Queensland 4064, Australia

John Wiley & Sons (Asia) Pte Ltd, 2 Clementi Loop #02-01,
Jin Xing Distripark, Singapore 129809

John Wiley & Sons (Canada) Ltd, 22 Worcester Road
Rexdale, Ontario, M9W 1L1, Canada

Library of Congress Cataloging-in-Publication Data

Norris, Mark.
    Designing the total area network: intranets, VPN's, and
enterprise networks explained / Mark Norris, Steve Pretty.
        p.   cm.   — (Wiley-BT series)
    Includes bibliographical references and index.
    ISBN 0 471-85195-7 (alk. paper)
        1. Wide area networks (Computer networks 2. Extranets (Computer
networks) 3. Business enterprises—Computer networks. I. Pretty,
Steve. II. Title. III. Series.
TK5105.87.N67 1999
        004.67—dc21   99-37649
                                                       CIP

British Library Cataloguing in Publication Data

A catalogue record for this book is available from the British Library

ISBN 0-471-85195-7

Typeset in 10/12pt Palatino by Vision Typesetting, Manchester

Printed and bound by Antony Rowe Ltd, Eastbourne

This book is printed on acid-free paper responsibly manufactured from sustainable forestry, in which at least
two trees are planted for each one used for paper production.

Contents

Preface

The need to communicate has never been greater. Many companies now rely on their internal networks for the fast and effective communication essential to their operation; and with an ever widening distribution of both information and business interests, the demands on these enterprise networks have risen. There is, for instance, an increasing demand to connect internal networks to those of external customers and suppliers—so called extranets or Community of Interest Networks (COINs). For these reasons, it is now widely acknowledged that the quality of an enterprise network—the communications infrastructure that supports an organisation—is a key competitive differentiator. But how do you ensure that you can buy, build or have access to the right enterprise network?

This is the question answered in this book. Effective and practical design techniques for enterprise networks are explained in detail and are illustrated with real examples. In addition to detailing what should be done, the text also covers what should *not* be done. The perils and pitfalls that pervade the high-tech, fast-moving world of communications are highlighted. Using the accumulated experience of many people, many projects and many networks, an extensive set of checks, balances, guidelines and recommendations are given. Because the book is intended for people in the workplace who need help and advice implementing an enterprise network it is deliberately biased to the (and, indeed needs to be) pragmatic and not the academic. Hence, it builds on a body of established knowledge of network technology—in particular, *Total Area Networking* already published in the Wiley/BT series.

This book has two unique features:

- It deals with the real issues and delves beneath the idealised facade of a complex subject, into the 'blood and guts' issues;

- it focuses on the activity of design—the essential skill of planning and assembling network technology to achieve the desired result.

Care has been taken to abstract from everyday technical complexity and make the key design issues clear and accessible. So the focus is on practical application rather than any particular technology in detail. That said, we have taken care to include real-world components in the case studies we present. The book is intended for a wide range of readers as in the following:

- An essential reference for the designers of enterprise networks. They will benefit from the concise presentation of key issues and the distilled wisdom contained in the checklists and guidelines.

- Recommended reading for those engaged in the purchase, management, planning and implementation of enterprise networks. This book provides the broad technical understanding required to ask the right questions, set viable plans and avoid expensive investment and deployment mistakes.

- A valuable professional updating guide for those who need to appreciate the practical subtleties of a complex subject.

- A useful text for introducing the practical application of core knowledge for final year and postgraduate students in computer science, electrical engineering and telecommunications courses.

The information age is no abstract concept—it is alive and evident in much of the latest network technology. Those who have the know-how to harness this technology will be able to work faster and more effectively. Those who do not adapt will cease to compete sooner, rather than later. Understanding the technologies central to the information age and how you use them to support your needs is, therefore, vital. It is important not only to know what is possible, but also what is not. What works and what does not. And, most important of all, how you maximise your own chances of success.

Mark Norris

and

Steve Pretty

About the Authors

Steve Pretty has 19 years' experience with BT in the data networking industry. After graduation, he spent 5 years designing X.25 packet switching hardware and software at BT Labs. He then moved on to design and implement data networks for BT customers, in both the financial and retail sectors. Networks for Electronic Funds Transfer were a key specialism during this period. When BT launched its Frame Relay service, Steve was one of the first engineers to get involved with supplying router based LAN interconnect networks to customers. He subsequently worked to develop the design methodologies and standards used today within BT Syncordia Solutions for the design and implementation of Enterprise Networks for customers. Steve is a Member of the British Computer Society and a Chartered Engineer. In his free time, Steve is a keen horseman, enjoying both dressage and cross country riding.

Mark Norris is an independent consultant with more than 20 years' experience in software development, computer networks and telecommunications systems. He has managed dozens of projects to completion from the small to the multi-million pound, multi-site and has worked for periods in Australia and Japan. He has published widely over the last ten years with a number of books on software engineering, computing, technology management, telecommunications and network technologies. He lectures on network and computing issues, has contributed to references such as Encarta, is a visiting professor at the University of Ulster and is a fellow of the IEE. Mark plays a mean game of squash but tends not to mix this with any form of networking. Mark can be contacted at mnorris@iee.org

Acknowledgements

The authors would like to thank a number of people whose help and co-operation have been invaluable. To those kind individuals who volunteered to review early drafts: Ed Smith, Chris Bilton, and John Atkins. Their observations, guidance and constructive criticism were always valuable (if, on occasions, painful) and have done much to add authority and balance to the final product.

Thanks are also due to our many friends and colleagues in the telecommunication and computing industries whose experience, advice and inside knowledge has been invaluable.

We have acquired our data network know-how over many years, through working with many individuals. Those deserving special thanks include a number from BT Labs; Paul Hitchen, Adrian Smith, Kevin Blakey, Steve Forrester and from Syncordia Solutions; Neil Taylor, Carlo Toon, Eoin McDonnell, Kevin McKay, Steve Tanner and Mick Deacy. And finally, we must thank the many customers we have worked with through the years, who have trusted us with their business and allowed us to hone our skills in the real world.

A Users Guide to this Book

Modern networks are complex. To design them properly requires a mix of skill, experience and dedication. This is something that you cannot readily impart, so you need some sort of preview and guide to what follows. This is it.

When we started this project we had an aim of giving a straightforward account of a subject often wrapped in baffling jargon and unnecessary complexity. After all, each of us had spent many years making sense of network systems and wanted to spare others from this subtle form of torture. The result is not the elegant and prescriptive manual we had in mind, though. In reality, the subject matter forbids this.

What we *have* ended up with is more of a 'cookbook'. Some sections give hints and tips, some give tried and tested recipes, some simply explain a particular piece of technology or fact of networked life. As with most cookbooks, we have tried our best to cater for a reasonably wide range of tastes. So some of the contents are aimed at the cordon bleu chef, some at the trainee commis chef and some at the enthusiast.

To help you select a suitable path through the full text, here is our summary of the delights that we think each chapter holds.

	Technical content	Experience content	Relevance to designers	Relevance to users	Relevance to managers
Chapter 1	**	***	**	****	****
Chapter 2	***	**	***	***	***
Chapter 3	**	*****	*	**	*****
Chapter 4	**	****	***	****	****
Chapter 5	****	****	*****	**	**
Chapter 6	*****	****	*****	**	**
Chapter 7	*****	****	*****	***	****
Chapter 8	***	****	*****	***	***
Chapter 9	****	***	***	***	*****
Chapter 10	****	*****	*****	**	**
Chapter 11	**	***	**	***	***
Appendix	**	***	***	****	****
Glossary	*	*	**	***	***

1

The Enterprise

There is nothing more difficult than to lead in the introduction of a new order

Niccolo Machiavelli

Network design is complex: part science, part craft. In many ways it is the same as any other form of design in that you establish the requirements, find the relevant components, fit them together, integrate the whole thing, test that it works and then release it for general use.

There is a bit more to effective network design than that, though. For a start, it is far from straightforward to articulate the requirements for a network, even one that is intended to support a few dozen people in a single office. A considerable amount of skill and know-how is needed to translate the wishes of a group of users into hard network plans ('hard' in the sense that they are expected to provide security, functionality, flexibility and low cost of ownership for a reasonable initial outlay). When the network being constructed is a core part of an organisation's infrastructure, an already difficult task becomes more of a challenge.

The basis of this challenge lies in the complexity of modern networks—and in the high expectations of the people that use them. In principle, it is possible to provide a user with speedy access to any information, whenever they want it, wherever they are, in a range of formats. In practice, though, there is a considerable amount of work to be done to deliver a modest subset of this. Careful and informed planning and design are required to satisfy the needs of the information worker.

Even when a satisfactory design is in place, it still has to be realised. Again, this is not a trivial task as the components that make up modern networks are not simple items. They are usually complex, state-of-the-art pieces of technology that need to be understood individually before they are used collectively.

Finally, when these complex components are all connected together, the way in which they will actually interwork is by no means predictable.

In short, you do not put effective networks together by luck—there are techniques to be used, guidelines to be followed, lessons of experience to be applied. And even then, there are no real guarantees. All is not gloom, though. There are plenty of examples of networks that provide competitive advantage, enrich their user's working days, even contribute to the environment by reducing unnecessary travel.

There are also examples of ineffective or unused networks that have soaked up time and money for little or no return. We assume that you do not want to end up in this category—so this book aims to put together the techniques, awareness and understanding that are needed to build practical information networks.

Our approach to explaining the design of real enterprise networks is layered. We work through first principles of networks and simple design techniques up to the grubby practical details of large-scale networks that use the most powerful technology. Along the way, common experiences of some of the real-world facts of network design are used to illustrate key points and to provide useful guidelines for the reader.

In working through increasingly lifelike design examples, we will highlight some of the practical concerns that the network designer has to face. For instance, how will they know they have finished, what are the best (indeed, possible) options, how can they make common-sense decisions (without divine intervention or the passage of years of study), should they believe what vendors tell them, how future proof will it be and how flexible must it be?

It should become clear in the course of these examples that real networks are very diverse, multi-faceted and inconsistent. Nonetheless, there are some fairly standard principles, ideas and issues that underpin the design of all communication networks—great and small. A general principle adopted throughout this text is a well established one, borrowed from Machiavelli. One of his quotes opens this chapter as it reflects our philosophy that complex problems can be understood, and subsequently solved, by examining the experiences of others who have faced similar problems in the past (Jay 1987).

Much of the advice and guidance included here is based on the accumulated experience of many network designers over the years. Technology may charge ahead, fashions may change and new requirements, capabilities and techniques may appear but some fundamental rules endure. We cover some of these basic principles later on.

In this chapter, though, we are concerned with building the broad picture. The aim is to outline the fundamental concepts and the key issues that designers need to know about—and this covers both technical and organisational matters. The former are required to deliver effective solutions, the latter to ensure that those solutions are appropriate to the customer.

The next chapter starts to relate this to practice by challenging the reader to design, from first principles, a communication network that supports the operation of a modern business. Where should they start? And what are the key issues in providing a fit for purpose solution?

Before embarking on this, though, we start by considering the roles and motivations of those involved. This gives us some of the background required to answer some of the above questions. After all, networks are put in place to serve people—and understanding the needs of those people is a vital first step in understanding how their needs can be met.

1.1   THE CREW

A network is installed by one group of people for the benefit of another group of people. Each of these groups has their own set of needs, which can be translated into measures of success for the network installation. As a whole, the groups form an enterprise that should receive some net benefit from the network.

In this section, we consider who the people involved in the enterprise are and what their roles, responsibilities and motivations should be. The real importance of doing this lies in the fact that each of the major roles has an influence—some direct (i.e. they make technical decisions, determine strategy or hold budgetary control), others indirect (i.e. they perform functions which the network will support or they enforce regulatory control over it). Here is a brief description of the skeleton crew—the minimum set of viewpoints that needs to be taken into account.

The IT director

The IT director is continually being pressed by the competition to be more efficient, by his or her peers to support an ever changing organisation and by users for greater convenience, speed and functionality.

The IT director is increasingly under the company spotlight as the information side of the business becomes more central to the organisation's profit, well-being and ultimate survival. More often than not, the pressure from the board is to outsource.

In some cases the IT director needs to support a competitive business that can operate on a global scale. He or she is attracted by the benefits of global presence, local knowledge so that things get done faster, cheaper, better (Cook *et al.* 1993).

The IT director wants an accessible window on enterprise network supply and operation and needs to get a grip on the information flow through the organisation.

The users

With an ever increasing need for information systems of which telecommunication networks play a vital constituent role, users want a reliable, hassle-free network that allows them to go about their job without getting involved in the technology that supports it.

Users want an easier life, a more rewarding job with less mundane administration and more creative challenge. They want fast access to the data they need, irrespective of location and original format. With an increasing trend towards virtual, distributed and flexible teams, they also want to feel that they belong (Brotchie 1991; Drucker 1988).

Users are the ultimate arbiters of a network's quality. A lot of time and money can be wasted if they reject the technology. It is key to involve this viewpoint as early as possible for two main reasons. The first is that the effectiveness of the end result depends as much on the use that this group make of the network as on its raw capabilities. Second, is the observation that behaviour always changes with the introduction of new technology. If the network is designed without user involvement, it is unlikely to function as intended when finally delivered for their use.

The IT professionals

The IT professionals need to understand the converging technologies of communications and computing so that they can provide the distributed information systems that are seen by the business gurus (Handy 1990; Porter 1986) as the competitive differentiator of the next millennium.

IT professionals want to be seen by their bosses (both line and contract) as a vital resource that cannot be dispensed with. They want to be seen as professionals committed to delivering the solutions that the business needs.

They also want a well-defined and ordered job with the satisfaction of constructing something that they are proud of, helped by rules of thumb with checklists. They would also like to enhance their future prospects by developing indispensable expertise and know-how.

IT professionals hold the key to what is practically possible—they are the experts who provide the various options for meeting the IT directors requirements. Sometimes they work for the IT director. Increasingly, they are contracted in to design, install or operate the network.

The suppliers

Suppliers have traditionally come in two flavours: network operators such as MCI, France Telecom, NTT and BT (for bandwidth) and computer suppliers such as Nortel, Cisco, IBM and Alcatel (for switches and the like).

Individually, they want to stay in business, so they need a market for their particular product lines. Increasingly, they want partnerships—with one another, to provide 'total solutions' for their customers and with customers themselves to provide 'facilities management'.

Collectively, they want an educated marketplace that can articulate needs and so guide their strategic developments.

Suppliers contribute to this aim by providing the majority of the technology push that has traditionally shaped the information network market.

The regulators and the law

It should not be forgotten that there are external factors that can have significant impact on network design.

The regulator is distinguished from all of the above in that they tend to be national rather than global (although export rules and trade agreements are changing this). As national players, they tend to focus their attention on basic services, rather than the enhanced or value added services.

One caveat that should be included here is that with one bit of digital information much like every other bit, how would any regulator know that their vision of the public good is not being subverted. The answer is that they do not (and cannot), so they focus instead on service offerings and network-type regulation. In terms of content (at least for general purpose media, such as the World Wide Web), it is unlikely that there will be a global regulator or 'bit police', and it is the customer who decides which bit they want and which bit they do not want.

The law is increasingly constraining use and design of networks. An example of this in the UK is the Data Protection Act, which can place limits on what information can be accessed over networks, who can access it and how it is conveyed (i.e whether encryption will be required).

So much for the various interested parties. They are not necessarily all relevant to every network but they each have something to gain from a deeper understanding of how enterprise networks are created (or they have something to lose by *not* understanding it!). For now, we settle on one of the above roles: the IT professionals who can delight or disappoint users, make or break the IT director.

1.2   THE MISSION

Modern organisations need to function faster, better and more efficiently to stay competitive. Increasingly, this entails the marshalling of resources that are distributed across a country or even across the world (Ohmae

1990). (The European Airbus consortia manages the construction of planes despite being distributed across four countries. Many finance and communication companies deal with transactions that involve tens of countries.)

Paper, fax and telephone are no longer sufficient. There is a clear need for computer and communication systems that allow people to work together at a distance as easily as if they were in the same office. The flexibility and suitability of these systems are seen by many as a vital differentiator in today's business world (Rudge 1993).

So, it is more and more in the interests of any high-tech organisation to ensure that their operational needs are satisfied by the computing and communication resources they pay for. And this means that they should provide a uniform and flexible backbone that takes the distance out of information (Sproull and Keisler 1991). This goal is embodied in the concept of the enterprise network—a combination of computing and communication resources that match the current operational requirements of an organisation . . . and can readily be recast to meet their future needs.

Let us dwell for a few minutes on what we mean by the term 'enterprise network' and what the implications of having one are. First, a definition: an enterprise network is an organisational resource that supports the transfer and processing of information across the full breadth of that organisation's operations. It should provide any client on the network with the services they require, irrespective of the physical location of those services (Goodman and Abel 1987). The concept of the enterprise network is illustrated in Figure 1.1—the client on the right of the picture gets the file and formatting services that have been built into this enterprise network (Hus and Lockwood 1993).

There are a few important points that are worth making about our definition of an enterprise network:

- The network should appear to both users and management as a single, homogeneous resource—irrespective of the actual variety of components from which it is built and breadth of physical coverage.

- The network is configured to directly support the specific operating procedures and processes—the user should be able to concentrate on doing their job, rather than having to drive a recalcitrant network.

- The network, although not necessarily owned by the organisation, should have defined management standards. It should perform to particular levels of performance, availability and reliability.

There is not much point in having an enterprise network unless you are going to capitalise on it. So what do you do with one—what use is it to you and what benefit do you get from acquiring one?

Many companies are achieving success by operating virtual teams. For example, John Sculley chairman of Apple Computers, predicts that tens of thousands of virtual organisations will come into place over the next 10

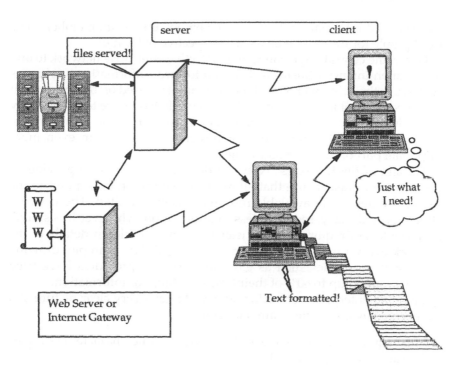

Figure 1.1   An Abstract View of an Enterprise Network

years These are groups of people who are assigned, irrespective of physical location or organisational structure, to a particular task or project. It is the role of the enterprise network to support the team; it provides the infrastructure for the virtual team to operate effectively.

So what? you may well ask. On the face of it, distributing people and jobs causes organisational headaches, inhibits teamwork and is traditionally frowned on as a means of getting the job done as fast and as well as you would like. Do virtual teams actually buy you anything, then? The answer (based on reported results) is yes! When it comes to information intensive work, they pay dividends and have been shown to yield:

- improved customer service (in 85% of cases)

- cost reduction (75%)

- improved quality (70%)

- faster delivery (80%).

Properly implemented virtual teams have, almost without exception, reported overall benefit. A survey, carried out in the UK by the government and industry trade organisations showed that 98% of the 356 organisations surveyed could identify substantial benefits. The key to success has been the increased information flow between empowered team members. The right people for the job can be used, irrespective of location.

Hence, the central importance of an effective enterprise network to any information-intensive business. The shift towards information working is striking. In the USA, over 50% of the job market is based on the production and processing of information, and this is predicted to rise to over 70% by 2015 (Swyt 1988). The use of networks to support these new ways of working is going to grow, so their design will become an even more important part of the information age.

The aim of the Enterprise Network designer should be to provide an information infrastructure that allows an organisation to respond to the market by deploying available resources, wherever sited. The network should provide easy, secure access to information, wherever it is stored, and should not require the end user to learn and relearn details of the network every time that something is changed. For example, drawing office staff in utilities such as gas and electricity providers have been shown to spend up to 60% of their time searching for information.

To get an enterprise network, you need to get a very clear picture of a number of key issues, the main ones being:

- user requirements (services to be carried, traffic characteristics, performance, throughput etc.);

- enterprise needs (how much this network is going to cost, how resilient and expandable it needs to be, how it fits with existing network resources);

- a way of working (a systematic method for combining user and enterprise needs and choosing, integrating, dimensioning and customising appropriate kit).

By the end of this book, we hope to have established a clear, common view of what these key issues are. And not just the technology but also the dynamics of their development, the position they hold within an organisation and their (inevitable) evolution.

This section has peered up at the stars and envisaged a world in which the user has complete command, can go where they want (at least in information terms) and can do what they want (that is, they have all the processing facilities they need) to produce the required result. But getting there is not easy, and we really do need to start the journey from ground level. Life is not perfect. You cannot just go out and buy a ready wrapped enterprise network. It is not a commodity purchase and all parties, customer included, have to do a fair amount of work to get an effective network (Norris 1995). The rest of this book illustrates this point, and it is a vital one to appreciate as expectations that a 'shrink-wrapped' network that will suit all needs are invariably dashed.

For now, let us put the attractive ideals of virtual organisations and information working to one side and start looking at practicalities. What do we need to do to achieve the ideals—what does an enterprise network consist of and how do you go about building one (or, more often, getting one built) that will suit your particular needs.

1.3   WHAT IS A NETWORK?

At its simplest, a network is no more than a set of connected entities, and in this particular instance, we are talking about communicating entities. Their function is to carry voice, data and video traffic so that end users can access, process, send and store the information that they work with.

In an age when information workers are becoming more and more important to the way organisations operate, networks to support them are becoming even more crucial. So a reliable, fast, effective network that supports the organisation's business—an enterprise network—is a core facility. If it is ineffective, business is lost. If it goes wrong, sometimes, the organisation is lost.

It is not difficult to illustrate the way in which information is supplanting tangible product. Even such everyday activities as booking into a hotel have become, in essence, no more than a transfer of information between two parties. In this instance, the hotel owner provides an access code for a room in exchange for the prospective resident's credit card number. The two parties need never meet and no tangible product or currency is exchanged. The transaction is effected with a network-connected card reader taking the buyer's details and a computer screen returning the required code to the user.

There are many more examples that illustrate the importance of pure information—some in everyday public use for example, electronic funds transfer, teleshopping, email and websurfing. And there are others in specialised areas of business for example, networked computer-aided design, electronic news gathering etc.

These sorts of applications can only happen if there is some means of transporting the information from one place to another, so that it can be appropriately processed, authorised and acted upon. To do this—at least over any sort of distance—you need a network.

A typical network looks something like the one outlined in Figure 1.2 (this is a fairly abstract view. A more 'nuts and bolts' picture of a typical network is given later, in Chapter 10). In reality, enterprise networks are composed of many different piece parts—the range of network technology is so wide that no single supplier can deliver a cost-effective solution to every user requirement. So the abstract symbols described, for instance, as 'multiplexers' and 'database' actually hide a whole host of technical and operational issues.

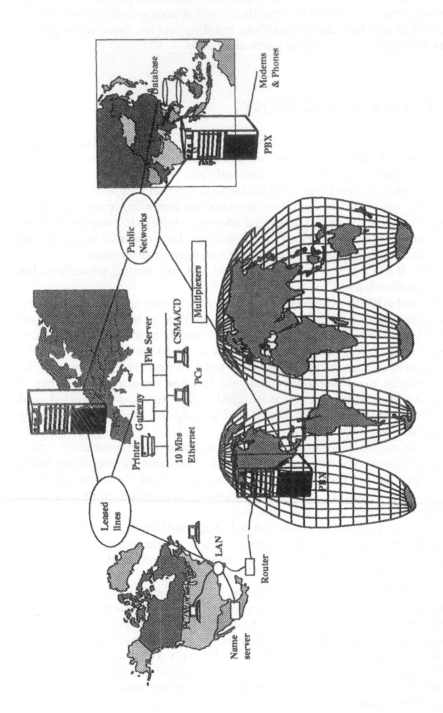

Figure 1.2 A Typical Enterprise Network

Real enterprise networks—those that allow an organisation to operate in the information business on a global scale—defy simple representation.

There are a few general observations based around the above figure that should be made. These are described below.

* *Management*
First, an enterprise network is not a static entity—it is in a state of continuous change. Some of these changes are intentional. For instance, new sites may be added, new facilities may be put on-line and routes may be configured to improve speed or access. There is another category of (less desirable) changes. These are the equipment failures, line outages and data corruption that afflict any computing or communication enterprise (and it is worth bearing in mind that enterprise networks both benefit from and suffer the woes of two, traditionally separate, areas of technology: computing and communication). For a network to provide organisation-wide support, it must be managed as a single resource and administered to meet the needs of its users.

* *Complexity*
The second point is that the physical representation of the network given above is but one of several viewpoints. As well as having physical and logical connections between sites, you also need compatible applications, common data formats, recognisable network names and addresses etc., and each one is vital to satisfactory operation. To carry out any viable sort of information work, you need to integrate a whole stack of protocols, applications and data links. However, this can all end in tears (or, looking at the layout of many networks, tiers) if incompatibility or end-to-end processing delay creeps in.

* *Security*
The final point here is that valuable resources need to be protected. If a network spans the world, it is open to abuse from anyone, anywhere. In truth, even a local network is open to global hacking these days. With information so valuable, security—both physical and logical—is paramount. The level of security needs to be balanced against the flexibility and ease of use seen by the user. Issues of data encryption, access control etc. will be visited later.

In general, though, these issues are containable—each one is covered in more detail later in this text. In practice, the composite parts of many networks work reasonably well with each other to provide the desired range of facilities—mail, telephony, file transfer. They continue to do so often enough and quickly enough to satisfy operational demands, but it is not easy—scale and complexity can conspire to defeat the unwary.

To give some idea of practical complexity, in the networks of large companies the average number of suppliers is typically between 16 and

20. Real enterprise networks do not have just three or four switching points, they can have hundreds, even thousands.

To make things even more difficult, each of the suppliers of network equipment usually offers a sophisticated product that needs to be understood if it is to be used to its best advantage and to work in harmony with the other components.

The network illustrated above will involve components from a variety of sources which might well include many versions and variants of the following:

- private exchanges, local area networks, gateways, bridges and routers, multiplexers from network equipment vendors;

- telephones, leased circuits and managed data networks from public network operators;

- computers, communications software, communications controllers and terminals from computer suppliers.

The enterprise network combines all of these elements; and not as a set of neatly organised boxes. The various elements of the enterprise network have evolved over the years and the next two sections trace some of the relevant history and background from two viewpoints. The first is the fairly stable telecommunications viewpoint, the second the fast moving world of networked computers. Each has its own concerns, dynamics and ways of doing things. They cannot be kept in isolation, though, as both have a significant role and contribution to the enterprise network.

Telecommunications

This means, literally, communications over a distance. In everyday language it now means the transmission of not just words but also sounds, pictures, or data in the form of electronic signals or impulses. Over the last 100 years, the telephone has come to be recognised as the most familiar form of telecommunications. More recently, voice telephony has been supplemented by a range of computer based telecommunication services. These have become popular through the Internet and World Wide Web, which provide many people with the means to exchange huge amounts of information.

Nowadays, it is taken for granted that by pressing a few buttons we can talk to family friends or business associates across the world. The technology that has led to mankind's largest and most complex creation—the telephone network—has been steadily refined over the last 100 years or so. Peter Huber refers to the hierarchically organised global telephone network as having 'the solidity, permanence and inflexibility of the Great Pyramid of

Cheops, which on paper it resembled' (Huber 1997). An assett, undoubtedly, but not the whole answer.

The modern telephone network can be viewed as a globally distributed machine that operates as a single resource. Much of this machine is comprised of interconnected computers and other intelligent devices. The network that most people use to carry voice traffic can also be used to transfer data in the form of pictures, text, and video images.

There are several ways of carrying information between senders and users. The options chosen should reflect the type of communication that is required. For instance, human beings are very tolerant of noise and transmission errors when they talk to each other. They are, however, very sensitive to delay. Computers have the reverse characteristics, being tolerant of delay, less so of errors. Here are some of the main concepts that underpin telecommunications networks.

- *Analogue and digital networks*
  Early telecommunications were analogue; they used continuously variable signals to convey information. The quality of speech across analogue networks was determined by the amount of the speech spectrum that could be carried. Around 3 kHz was accepted as a reasonable compromise of cost and quality. The analogue voice networks of the 1970s have migrated, piece by piece, towards today's digital networks. This started with long distance and inter-exchange connections and, more recently has extended into the local network. Many digital data networking solutions, for instance ISDN, have grown out of the digital voice service. The fact that such services are based on multiples of 64 kbps or 56 kbps (the digital equivalent of 3 kHz bears witness to the roots of data networks in voice technology!).

  Computer communications, which are based on discrete, digital signals can use analogue connections but are limited. The original drive for introducing digital networks was cost—a happy consequence that they are ideally suited to computer networking. The capacity of digital networks has grown very rapidly (e.g. from 64 kbps in the mid 1980s to 10 Mbps and more by the mid 1990s 155 Mbps and higher in the late 1990's) and they can carry a mix of voice, data, text and pictures—one bit is very much like another!

- *Circuit switching and packet switching*
  The distinguishing feature of circuit switching is that an end-to-end connection is set up between the communicating parties, and is maintained until the communication is complete. The Public Switched Telephone Network (PSTN) is a familiar example of a circuit-switched network. Indeed, it is so familiar that many people are not aware that there are other ways of doing things.

  Communication between computers, or between computers and

terminals, involves the transfer of data in blocks rather than continuous data streams. Packet switching exploits the fact that data blocks can be transferred between terminals without setting up a fixed end-to-end connection through the network. Instead they are transmitted on a link-by-link basis, being stored temporarily at each switch *en route* where they queue for transmission on an appropriate outgoing link. Routing decisions are based on addressing information contained in a 'header' appended to the front of each data block. The term 'packet' refers to the header plus data block.

The Internet is based around packet technology. It is the size of the Internet that enables the packets that support all applications to find a variety of alternative paths to their destination (and its heavy usage that, on occasions, accounts for the time taken to do so).

- *Congestion and blocking*
  In a packet-switched network, packets compete dynamically for the network's resources (buffer storage, processing power, transmission capacity). A switch accepts a packet from a terminal largely in ignorance of what resources the network will have available to handle it. There is always the possibility therefore that a network will admit more traffic than it can actually carry with a corresponding degradation in service. Controls are therefore needed to ensure that such congestion does not arise too often and that the network recovers gracefully when it does.

  In a circuit-switched network the competition for resources takes the form of blocking. This means that one user's call may prevent another user from getting access. Since the circuit is reserved by the user— irrespective of what they send—for the duration of their call, no-one else has any form of access until the call is cleared. Traditional circuit-switched networks are designed to balance the amount of equipment deployed against a reasonable level of access for the user's of that network.

- *Performance*
  A circuit-switched network, such as the PSTN, provides end-to-end connections on demand, provided the necessary network resources are available. The connection's end-to-end delay is usually small and always constant, and other users cannot interfere with the quality of communication.

  In contrast, in a packet-switched network packets queue for transmission at each switch. The cross-network delay is therefore variable, depending as it does on the volume of traffic encountered *en route*.

The Internet (a network of computer networks that share a common set of protocols and address space) has grown over the last 25 years and now links together around many tens of millions of people world-wide, mostly over the telephone network. Even the White House in the US and Houses of Parliament in the UK have pages on the Internet (these can be accessed,

respectively, via the World Wide Web at URLs http://www.whitehouse.gov
and http://www.open.gov.uk).

The traditional national network providers source much of the capacity
used to support the Internet. That said, the traditional telephone networks
are having to evolve in order to cope with the traffic profiles caused by
internet connections. In fact, they are major suppliers across the broad
range of data and voice communications. Over the last few years, they
have tended to expand beyond their traditional, national boundaries
(usually through alliances) so that they can meet the demands of the
growing enterprise network market. These alliances are far from stable—the
author's map of the way in which the telecommunications providers are
evolving to cover the three major world markets in the US, Europe and
Asia-Pacific regions was too volatile to place in a book!

The rationale that underpins these alliances is that they make it viable to
sell virtual private networks. These are attractive from the customer's
point of view in that they can hide much of the complexity in networks
with many nodes. This is illustrated below, where the addition of more
sites dramatically increases the number of router ports required. A virtual
private network offers the prospect of keeping the customer's problem to a
minimum. A potential problem is left within the 'cloud' administered by
the telecoms operator (often referred to as the Telco).

The cloud shown in Figure 1.3 represents the virtual network provision
by a third party—in effect, it provides a link for each node connected to the
cloud. This is an attractive way of building in room for growth. It should
not be treated as an infinitely adaptable panacea, though and some of the
perils of 'cloudism' are explored in Chapter 3.

So much for the traditional area of telecommunications—the essential
fabric of a network. But there is yet more to be done before the whole thing
is ready for use. Many of the applications that operate over networks
(telephony aside) are sourced by the computing industry. These suppliers
have different concerns and a different way of doing things.

Computer networks

These are a central part of the Information Age. The popular adoption of
the personal computer (PC) and the local area network (LAN) during the
1980s has led to widespread sharing of information among groups of
computers and their users. It is taken for granted these days, that you can
access information on a distant database, download an application from
overseas, send a message to a friend in a different country and share files
with a colleague, all from your own computer.

The networks that allow all this to be done so easily are sophisticated
and complex entities. They rely for their effectiveness on many co-operating

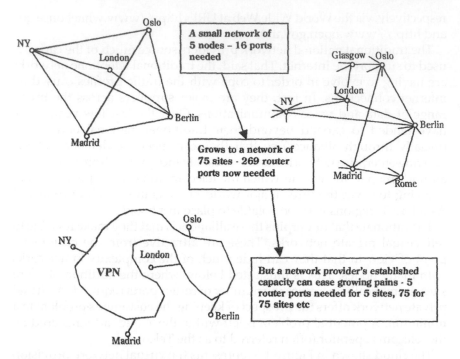

Figure 1.3 An Illustration of the Appeal of the Virtual Private Network

components. The design and deployment of the world-wide computer network that now spans the world can be viewed as one of the wonders of the modern world. It has been made possible only through the ingenuity, persistence and co-operation of many people over many years.

As recently as the 1970s, computers were expensive and fragile machines that had to be looked after by specialists and kept in a controlled environment. They could be used either by plugging a terminal directly or by using a phone line and modem to gain access from a distance. Because of the cost of computers, they tended to be centralised resources to which the user had to arrange their own access. During this time, computer bureaux (organisations that offered access time on a mainframe computer) flourished. Computer networks during this era were not commercially available. Even so, one of the most significant developments to shape the modern world of technology was initiated at this time. This was experimentation by the US defense department in distributing computer resources to provide resilience against failure. This work is now known to many people in the form of the Internet.

One of the most dramatic events in computer networking has been the introduction and rapid growth of the local area network (LAN). As the name suggests, this is a means of connecting a number of computing elements together. At the simplest level, a LAN provides no more than a shared medium (e.g. a coaxial cable to which all computers, printers etc. are connected) along with a set of rules that govern the access to that medium. The most widely used LAN, Ethernet, uses a mechanism called Carrier Sense Multiple Access with Collision Detection (CSMA/CD). This means that every connected device can only use the cable once it has established that no other device is using it. If there is contention, the device looking for a connection backs off and tries again later. The Ethernet transfers data at 10 Mbits/s, fast enough to make the distance between devices insignificant. They appear to be connected directly to their destination. Another popular LAN, the Token Ring, achieves the same end (as its name suggests) by passing a token that grants access permission around a ring of connected devices.

Ethernet and Token Ring are just two general examples of LANs. There are many different layouts—bus, star, ring—and a number of different access protocols. Despite this variety, all LANs share the feature that they are limited in range (typically they cover one building) and are fast enough to make the connecting network invisible to the devices that use it.

In addition to providing shared access, modern LANs can also give users a wide range of sophisticated facilities. Management software packages are available to control the way in which devices are configured on the LAN, how users are administered, and how network resources are controlled. A widely adopted structure on local networks is to have a number of servers that are available to a (usually much greater) number of clients. The former, usually powerful computers, provide services such as print control, file sharing, database management, mail etc. to the latter, usually personal computers.

The facilities on most LANs are very powerful. Most organisations do not wish to have small isolated islands of computing facilities. They usually want to extend facilities over a wider area so that groups can work without having to be co-located. Routers and bridges are specialised devices that allow two or more LANs to be connected. The bridge is the more basic device and can only connect LANs of the same type. The router is a more intelligent component that can interconnect many different types of computer network.

Many large companies have corporate data networks that are founded on a collection of LANs and routers. From the user's point of view, this arrangement provides them with a physically diverse network that looks like one coherent resource.

At some point, it becomes impractical to extend a LAN any further. Physical limitations sometimes drive this but more often than not, there are more convenient or cheaper ways to extend a computer network (Williamson 1994). Two major components in most real computer networks are the public telephone and data network. These provide long distance links that extend a LAN into a wide area network.

Nearly all of the national network operators (e.g. DBP in Germany, BT in the UK) offer services for the interconnection of computer networks, referred to earlier as virtual private networks. These are based on a range of technologies, from simple, low speed data links that work over the public phone network through to sophisticated high-speed data services (such as Frame Relay and SMDS—Switched Multi-megabit Data Service) that are ideally suited to the interconnection of LANs. The high speed data services are usually referred to as broadband connections (IEEE 1992). It is widely anticipated that they will provide the necessary links between LANs that make the Information Superhighways a reality.

It would be easy to assume that computers will all be able to work together once they have broadband connection. But how do you get computers made by different manufacturers in different countries to work together across the world? Until recently, most computers were built with their own interfaces and were structured in their own unique way. A computer could talk to one of its own kind but would have difficulty communicating with a foreigner. There were only a privileged few with the time, knowledge and equipment to extract what they wanted from a variety of computing resources.

However, this situation improved rapidly through the 1980s. The level of commonality across different computers has now reached the stage where they can interwork effectively. This allows virtually anyone to use remote resources to good effect. There are several factors that have helped to contribute to this happy situation. The main ones are as follows.

- *Client–server*
  Instead of building computer systems as monolithic systems, there is now general agreement that they should be constructed as client–server systems. The client (e.g. a PC user) requests a service (e.g. printing) and the server (a LAN connected processor) provides it. This consensus view on the structure of a computer system means that there is a separation of previously bundled functions. The implementation details that flow from a simple concept go a long way to enabling all computers to be treated uniformly.

- *Object Technology*
  This is another philosophy on the way of building computer systems. It works from the premise that they should be built from well defined parts—objects which are encapsulated and are defined and implemented so that they can be treated as independent agents. The adoption of objects as a means of building computer systems has helped to allow interchangeability of parts.

- *Open Systems*
  This term covers the general aim of building computer systems that they can readily be interconnected, and hence distributed. In practice, open

systems are all about unbundling all the complexities of a computer system and using similar structures across different systems. This entails a mixture of standards (which tell the manufacturers what they should be doing) and consortia (groups of like minded people who help them to do it). The overall effect being that they can talk to each other.

The ultimate aim of all of the work in distributed systems is to allow anyone to buy computers from a number of different manufacturers, to site them wherever is convenient, to use broadband connections to link them and to operate them as one co-operating machine that takes full advantage of these fast connections.

The explosive success of the TCP/IP protocol suite, which has grown out of its use as the basis of the Internet, has done much to make the open systems dream a reality. TCP/IP has evolved over the years and come to dominate the marketplace.

Having fast computer networks built of machines that can talk to each other is not the end of the story. The spectres of the Information Superhighwayman and the Information Superroadworks have yet to be dealt with.

- *Security*
  With ever increasing amounts of important information being entrusted to ever more distributed computers, security becomes even more important. In a highly distributed system it would be all too easy for an informed superhighwayman to access confidential information without being seen. The Data Encryption Standard (DES) standard for protecting computer data, introduced in the late 1970s, has more recently been supplemented by 'public key' systems that allow users to easily scramble and unscramble their messages without a third party intruding.

- *Management*
  It is a full time job to keep a LAN operating properly. Keeping a computer network that is distributed across the world running smoothly takes the challenge of network management one leap further. The essential concepts for managing distributed and diverse networks have received a lot of attention over the last few years. There are now enough tools and standards for this important aspect of computer networks to allow many networks to be effectively supervised. The management of very large networks still requires hand-crafted solutions, though, with compromises having to be made between management overhead and level of control.

The challenge facing the network designer is the integration of these separate components into a coherent network in a cost-effective manner. The set of components used must not only interwork—it must also be readily operated and maintained as a single resource.

To meet this challenge, you need to go further than drawings of boxes with lines between them. The picture of our typical network does not show what capacity the links between elements are, how fast the switch has to work, what protocols the terminating equipment needs to understand etc. To answer these questions—and to get to the real meat of designing networks—you need to systematically match user needs with processing, transmission and switching capability. The remainder of this chapter illustrates some of the techniques for tackling this task. The next chapter will illustrate it all by going through a real(ish) design.

1.4   THE DESIGNER'S JOB

There are many challenges for the designer to meet. Here we give a brief resumé of the key quality issues that the designer needs to address and some of the main functions that they perform.

In general, the effectiveness of any design (or its 'quality' or 'fitness for purpose') can be measured against the following criteria:

- User needs. The ultimate purpose of an enterprise network is to provide those who use it with a means of doing their job faster, better and more easily. So the final test of suitability for purpose is the extent to which the network supports the applications that run over it. The remainder of the points here are, to a greater or lesser degree, aspects of user need.

- Cost. Not just the cost of purchasing and installing the new network, but also the cost of keeping it. As a rule of thumb, the cost of ownership of a software rich systems (and enterprise networks increasingly fall into this category) is twice its initial cost. So the cost of the design—its whole life cost—is an important consideration for the designer to bear in mind.

- Performance. As with cost, there are two sides to this quality attribute. First, the design of the network must allow transactions to be executed quickly enough to satisfy the user. This entails careful end to end checks for the range of applications being supported. The second performance issue is throughput—not only must the network be fast enough, it must also cope with the peak traffic volumes.

- Reliability. This is another end-to-end issue for the designer to consider. The perceived reliability of any network relies on all the elements that contribute to a user's session working at the same time. So, it may be that one or more network elements can go down without affecting the user. That would be a well-designed network. A poorly designed network may keel over every time that one particular element goes wrong. There is, of course, a trade-off between the level of reliability and cost.

- Availability. Some level of maintenance and administration is inevitable with any network. The extent to which this intrudes on the user is a measure of fitness for purpose.

- Expandability. Today's new system is tomorrow's legacy. Design for change is an increasingly important factor with organisations continually changing structure, procedures and working practices. A good design should be built to accommodate the incorporation of new services, dynamic capacity management and some means of integrating data from a range of sources.

- Manageability. It should be possible to keep as expensive and dynamic an asset as an enterprise network in good shape. This means that it must be designed to be managed. So each element of the network must be open to some standard form of interrogation from a network management system. The specialist area of network and service management is a complex one, considered in a later chapter.

The trouble with these oft-quoted benchmarks is that they are all rather difficult to quantify. Also, they are mostly retrospective with the real limits and thresholds only appearing once the network is installed and being used.

One of the biggest challenges facing the designer is to get some handle on complexity. The desirable features described above can only really be tackled if the overall design can be broken down in such a way as to allow some form of analysis. An example of how this can be achieved is shown in Figure 1.4. This illustrates the use of three views on a design—a technical view, a functional view and an enterprise views (Zachmann 1987). The idea of this is to get a handle on the overall view before diving into the nuts and bolts. There are other (similar) guidelines for systematically tackling the complexities of design. The use of one of the established guidelines is generally accepted as useful—the key point being that some organised format for structuring abstract thoughts is a good idea.

As soon as there is a reasonable overall picture, some level of detail can be added. Much of the subsequent text illustrates the way in which high-level requirements can be elaborated, checked and refined to yield a viable design that balances all the various viewpoints and qualities described here in an equitable way. In practice, such balance is sometimes not achievable. However, it is still important to have a systematic approach as a basis for justifying/defending choices that are taken. It is not a short story, though, as there are many problems and pitfalls that have to be dealt with.

Our overall aim in this book is to build up a practical guide for the designer to use and for other interested parties to understand. A good approach to design should allay some of the fears and pressures on the IT director, keep the users happy and give the suppliers a clear view of what they need to do. So the whole crew should be well served by what is to come.

1.5 ABOUT THIS BOOK

There are three separate parts to this book, each of which has a distinct

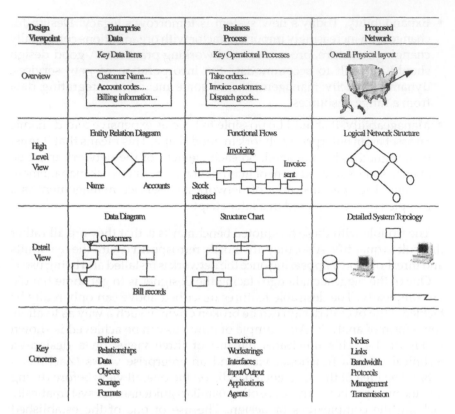

Design Viewpoint	Enterprise Data	Business Process	Proposed Network
Overview	Key Data Items Customer Name.... Account codes.... Billing information...	Key Operational Processes Take orders... Invoice customers... Dispatch goods...	Overall Physical layout
High Level View	Entity Relation Diagram Name Accounts	Functional Flows Invoicing Invoice sent Stock released	Logical Network Structure
Detail View	Data Diagram Customers Bill records	Structure Chart	Detailed System Topology
Key Concerns	Entities Relationships Data Objects Storage Formats	Functions Workstrings Interfaces Input/Output Applications Agents	Nodes Links Bandwidth Protocols Management Transmission

Figure 1.4 The Range of Issues that Face the Enterprise Network Designer

style and purpose. The first part (in the next chapter) is an idealised view of network design. It looks attractive, but in practice it simply does not work!

From this false dawn, we move on to build up a systematic and informed process for the designer, complete with the necessary checks, balances, rules of thumb and guidelines to tackle real network designs (see Chapters 3–9). The third part (Chapter 10) illustrates all this with an example drawn from the author's practical experience.

So, the next chapter starts to deliver something useful by going through an abstract view of network design. The key stages of design are introduced here and it is intended that, by the end of this chapter you will know how to design networks—at least in principle. However, there is a sting in the tail in Chapter 2. When we start to introduce a few of the real world problems that make networks tricky, we realise that design is complex and not something that can be done with the simple prescriptions.

Having established that a more detailed approach is needed, Chapters 3–9 add much of the practical flavour for real enterprise network design. In this section we cover requirements gathering and analysis, architectural design, top level design, the business case, procurement and supplier management, system testing and trials, implementation design, implementation, problem resolution, change management, quality assurance and planning for the future. By the end of this section it is intended that you should know how to design networks in practice, appreciate key practical challenges and have a comprehensive checklist for key issues.

After this section, move to the third part of the book with a practical illustration. Chapter 10 deals with the current and future enterprise network for a company called Data Bank. The example of Data Bank is an assembly of a number of real network projects—not identifiable but heavily based on reality. In this part of the book we aim to illustrate the application of the previous guidelines, provide some anecdotes to help lock the subject matter into the readers world and, hopefully some wisdom for guidance (we should say that much of the 'wisdom' is the product of many projects, many people's thoughts, efforts and deeds). Since the aim in this section is application, we point to the wealth of excellent references in this area for specifics rather than stop for explanation.

To close, we take a peek over the horizon at some of the future directions for enterprise networks. In particular, we consider the growth of multimedia applications, the possible advent of Asynchronous Transfer Mode (ATM) as the platform for multimedia networking and as the Platform for Total Area Networking (Atkins and Norris 1998). To link this with the 'here and now', we also say a little on the commercial use of the Internet and explain some migration strategies (i.e. ways of making legacy networks work—how to avoid millstones, and the use of Virtual Private Networks).

1.6 SUMMARY

In a world more and more dominated by communication and information, networks are key. They provide the nervous system that any information-intensive business needs to compete (Gilder 1994).

This chapter has introduced the notion of enterprise networks, and has discussed their inexorable rise, who has an interest in them and what they should, ideally, be designed to achieve.

In reality, though, the enterprise network that fulfils all of the needs of the information age is a rare beast. Most of us still use the telephone as our primary means of communication, strain our eyes reading distorted fax messages, travel to meetings, and struggle with long and convoluted electronic mail addresses before resorting to sending disks via surface mail because file transfers are too slow and tortuous.

There are two main messages here. The first is that enterprise networks are not optional—they will be an essential part of many organisation's operations. The second message is that the general state of understanding is poor and many networks just grow rather than being planned—networks are not a commodity purchase, they require skill, know-how and hard work to get right. The purpose of this book is to share our understanding of practical network design.

This chapter has made a start in explaining some of the basic commercial and technical issues. The chapters that follow are aimed at building a balanced and detailed guide for those who wish to take advantage.

REFERENCES

Atkins, J. and Norris, M. (1998) *Total Area Networking—ATM, IP, Frame Relay and SMDS explained*. John Wiley & Sons.

Brotchie, J. (1991) *Cities of the 21st Century*. Longman Cheshire.

Cook, P. *et al*. (1993) *Towards Local Globalisation*. UCL Press.

Drucker, P. (1988) The coming of the new organisations. *Harvard Business School Review* Jan/Feb, 45–53.

Gilder, G. (1994) *Telecosm*. Simon & Schuster.

Goodman, G. and Abel, M. (1987) Communication and collaboration: facilitating co-operative work through communication. *Office Technology and People*, **3**, 129–145.

Handy, C. (1990) *The Age of Unreason*. London Business Books.

Huber, P. (1997) *Law and Disorder in Cyberspace: Abolish the FCC and Let Common Law Rule the Telecosm*. Oxford University Press.

Hus, J. and Lockwood, T. (1993) Collaborative computing, *Byte*, March.

Strategic impact of broadband communications in insurance, publishing and healthcare (1992). *IEEE Journal on Selected Areas in Communication*, **10**, 9, December.

Jay, A. (1987) *Management and Machiavelli*. Hutchison Business Books.

Norris, M. (1995) *Survival in the Software Jungle*. Artech House Books.

Ohmae, K. (1990) *The Borderless World*. HarperCollins.

Porter, M. E. (1986) *Competition in Global Industries*. Harvard Business School Press.

Rudge, A. (1993) I'll be seeing you. *IEE Review*, November, 235–238.

Sproull, L. and Kiesler, S. (1991) Computers, networks and work. *Scientific American*, Issue 3, 265.

Swyt, D. (1988) The workforce of US manufacturing in the post-industrial era. *Technology Forecasting and Social Change*, **34**, 231–251.

Williamson, J. (1994) The LAN/WAN maze. *Global Telephony*, September, 20–31.

Zachmann, J. A. (1987) A framework for information systems architecture. *IBM Systems Journal*, **26**, 3.

2

The Basics of Network Design

Skill comes so slow and life so fast doth fly. We learn so little and forget so much

Sir John Davies

In this chapter, we build on the background material from the previous chapter and start to advise on how to approach the design of real networks. Some of the practical issues that the designer needs to deal with are introduced in this chapter. That is not to say that we have found a quick fix to a complex problem. Rather that this is a first step in the right direction.

A systematic design process, consisting of five stages is developed in this chapter. It is illustrated through the development of a realistic example. This does not show all of the options or routes open to the network designer, nor does it cover the full range of issues that the successful designer needs to deal with. Nonetheless, it does bring the strengths (and limitations) of a systematic procedure into sharp focus—an essential grounding for subsequent elaboration.

There are two reasons for diving in so deep, so early. The first is show that there are some aspects of network design that can be (and should be) codified, analysed and resolved through convergent thinking. Some useful and relevant techniques for supporting logical thought should emerge from the case study that unfolds in this chapter.

The second, and perhaps less honourable purpose is to set a dividing line between logical study and informed fire-fighting. Too much of real network design relies on the latter for it to be ignored. That is not to say that we give up any sort of ordered approach when the chaotic world intrudes. The next chapter takes the basic principles and overall structure built in this chapter and adds some judgement and guidance borne of experience.

By the end of this chapter, it should be clear what you need to do to design an enterprise network—in principle. There are many real world imperfections that mask this ideal, but our aim here is to introduce some basic structure and to set the scene for the real world lessons illustrated later.

Incidentally, it is assumed that many readers will be familiar with some or all of the technical jargon and concepts used in this chapter. If not we have appended an extensive glossary (and there are many good books available on the underlying technologies, some of which we will reference). That part of the book is intended for reference rather than being a part of the main storyline.

2.1   A SYSTEMATIC APPROACH

Let us say, for the purposes of illustration, that we have been asked to design a network to support the various operating divisions of a large organisation. This entails designing from scratch—there is nothing in place to start with, so we have a free hand to build what is most suitable for the job.

The first point to make in this chapter on practicalities is to explain what is meant by 'nothing in place to start with'. This does not mean that we start our design in isolation from everything else. Contrary to what many texts on the subject imply, there is no such thing as a 'green field' design. Even if we were completely overhauling an organisation's network, there are still established connection points that have to be considered, external systems that have to be interfaced to, sources and repositories of data to be accessed and essential processes that need to be supported.

So, the first thing we need to do is to establish the framework in which we are operating. This, along with the end result that the user wants to see, constitutes the vital information for building a full set of network require-ments. This is the first of a series of systematic steps that should be completed by the designer—establishing where they are going and where they are starting from.

A second point that needs to be made about practical network design is that it is not prescriptive. For reasons of observability and traceability, it should be carried out systematically, but ideas and options evolve as requirements and their fulfilment become clearer. By considering this need to iterate towards a solution, we can begin to give a naive, first draft outline of the main design stages. So design choices taken early on need to be examined, checked and tested. If a particular way of doing things proves too expensive, too slow, etc., then it may be necessary to backtrack to a point where different options are taken.

Either way, the key point of this section is that an often obscured history combined with an always unpredictable future make it essential to adopt a design approach that can cope with change. This means some level of procedure in the development of the design process.

2.2 FIVE STAGE NETWORK DESIGN

In view of the issues raised in the last section, we put forward here a process for network design that constitutes five basic steps. These are:

- Requirements—deriving a statement of what the customer wants of the network that is being built. Ideally this starts from their business need. The statement of requirements (which needs to be a joint effort of customer and supplier) should be detailed enough to allow proposed designs to be tested for acceptability. It should therefore contain quantitative as well as qualitative statements. In truth the former are usually notably thin on the ground!

- Architectural design—examining the various ways in which the requirements can be met and proposing (usually after presenting a short list to the customer) which option should be adopted. The resulting design should give a good idea of the type of technology to be used.

- Top-level design—filling in the detail from the previous stage. This demonstrates how the anticipated traffic volumes, types and mix can be carried on the outline design. It shows how the network will be implemented in terms of physical links and where the traffic flows will be. It also works out the required numbers and capacities of circuits, terminals, etc.

- Implementation design—focuses on the logical aspects, such as naming and addressing traffic routing policies. It checks that the service parameters in the statement of requirements are realisable and that the right capacity links are available between the right locations. It verifies that connections, ports and machines are identified, that the right versions of software are installed and a whole host of other practical issues are adequately covered.

- Business case—working out a likely cost for the network. This covers both capital and current expenditure and should cover secondary costs such as software and hardware upgrades, general maintenance, licences and contracted services.

If you are very fortunate, these five steps are sequential and result in a network design that, when implemented, does exactly what the customer asked for in the first place. More often, though, there are dead ends due to misguided early decisions, changing or missed requirements, etc. Also, you have to go back and start over at an earlier part in the process. So the above set of activities should really be seen as elements of an interactive lifecycle. If any one step fails, you need to go back to the last valid step.

One further point in the above process is that it covers only the design process. Issues such as change management and problem resolution have to be included in the delivery of any real network. These are covered later. For now, though, we will stay with the basics.

The main point in putting together a process such as this one is not that it gives any sort of guarantee of success but that it allows decisions and options to be traced. Not only does this demonstrate a professional approach, it also instils some confidence that the final solution is fit for purpose and has been adequately exercised prior to (usually very costly) implementation.

So much for the merits of doing things in an ordered way. Now for an example to illustrate what sort of things are actually involved. The case study used in the remainder of this section is based on an amalgam of several actual network designs It is intended to introduce the issues in designing a typical, real-life enterprise network.

Requirements

The first step is to work out where you are starting from and where you want to get to. In this instance, the network needs expressed by the customer are fairly specific as they have fairly rigid operating procedures that they need to maintain. The initial layout of the company's enterprise network is illustrated in Figure 2.1. Its limitations are fairly evident, the main problems being the isolation of the Berlin site and the limited connections between the main database in New York and its reference and archive relatives in London and Madrid, respectively.

The main reason for the mismatch between what exists and what is needed is historic. The current enterprise network was built up in a piecemeal fashion over about 10 years and it has only recently become evident, with increasing competition and rate of change, that it has serious operational limitations. These shortcomings are reflected in the stated customer requirements which, at the highest level, give several key criteria for the new network—that it must:

- completely connect the five main sites currently operated by the organisation;

- entail minimal changes to existing site LANs;

- provide a support infrastructure for the enterprise-wide mail system;

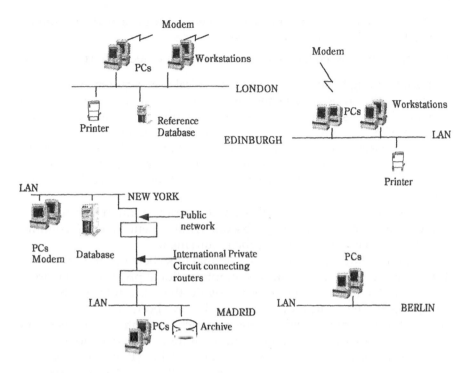

Figure 2.1 The Current Network

- allow common access to the master database, sited in New York;

- allow easy addition of a new site (probably in Oslo).

These high-level needs will need to be refined as we work though the design options. And even this small-scale design would take considerable time and space to analyse fully.

In view of this, it is worthwhile (and generally good practice) to build some user scenarios for the network. These provide some target against which the evolving design can be sanity checked, as opposed to exploring all possible options.

Some of the salient points that emerged from talking to prospective users of the new network were that they expected:

- all files to be downloaded to Madrid every night for storage;

- the main database in New York to be routinely accessed from both London and Edinburgh as an essential part of daily operations;

- guaranteed delivery times for particular transactions;

- administration and configuration of the network to be carried out in-house, by a team in Berlin with relevant skills;
- security to be assured for a small but important number of the transactions.

It is interesting to note that the above 'typical user requirements' contain no quantitative information—the third statement on guaranteed delivery conveys the user's desire to have a network with predictable performance and the fifth the feeling that some traffic should be protected. The lack of figures associated with these statements indicates that some requirements are based on intuition—part of the designer's job is to tease out what is an acceptable delivery time, and what level of security is really necessary.

Progress can be made from this point, bearing in mind that there are some issues for negotiation and subsequent refinement. Figure 2.2 outlines a first cut, naive solution to meet the requirements.

The figure shows a number of additional links between the company's LANs—each of the added lines is an international private leased circuit with a router at each end. These provide the basic total connectivity requirement. This same requirement could have been met in a number of different ways, for example by connecting each LAN into a public network. This would be shown as a set of connections into a network 'cloud' which provides all the required transmission and switching that is required. The perils of this design approach are discussed in the next chapter. However, this proposal seems reasonable as a first suggestion—something to start exploring in a little more depth.

Architectural design

The next thing to get straight here is what technology is the most appropriate to provide the new links shown in Figure 2.2. The first part of this is to consider the technical need. As with many enterprise networks, this concerns the information bottleneck that lies at the boundary between the local and wide area networks, as illustrated in Figure 2.3.

A typical local area network operates at speeds in the range 10–100 Mbps. They tend to cope well with the bursty traffic generated by most information workers. In this instance, the LANs on each of the sites are set up for client/server operation and the users need to carry out a lot of remote file accesses. These all demand high bandwidth for short periods of time. In contrast, the wide area network links (usually leased from the public telecom operator) tend to offer much less rapid rates and are oriented towards semi-permanent connection, rather than variable bandwidth. This interface is one that concerns many designers. Hence the key technical need—to get a closer match between the local and wide area facilities, thus improving the overall network performance.

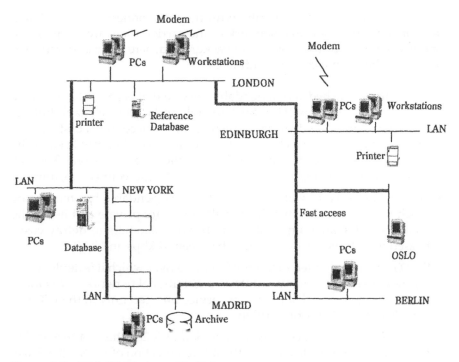

Figure 2.2 PC/LAN Clusters with Links

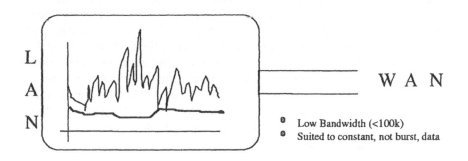

- Low Bandwidth (<100k)
- Suited to constant, not burst, data

- Bursty Data
- High Bandwidth (10 Mbps +)

Figure 2.3 The Information Bottleneck

Since we are dealing, primarily, with the interconnection of local area networks using a wide area network as the carrier, there are a number of options that could be used. By way of background, here is a brief description of each of the main technical candidates:

- X.25—an established and dependable service that is generally available at speeds up to 64 kbps. It is robust, reliable and widely available but has significant end-to-end delay (which is introduced by the store and forward nature of the technology). Prices are usually fairly keen as the technology is well established and the service is a public one. The low speed and delay make it poorly suited to LAN interconnection, though.

- Leased lines—again well established and dependable but tend to be much more expensive than X.25 as the links are private and hence the media is not shared amongst a number of customers to defray costs. Leased lines typically offer speeds between 64 kbps and 2 Mbps.

- SMDS—fast but limited in terms of the area over which it is deployed as it is not available in many countries and national networks are seldom connected. Very much a premium service but, certainly in the UK, is a surprisingly cost effective option.

- Frame Relay—often described as the 'Son of X.25', this is a reasonably fast service that is well suited to the interconnection of LANs. It is fairly widely deployed and often very reasonably priced. When it comes to international networks there is a fundamental difference between X.25 and Frame Relay—X.25 is primarily a set of networks run by national PTTs, widely interlinked by X.75 gateways. So there is very wide geographical coverage, but there is also an obligation to deal with many suppliers (although there are some exceptions such as Concert Packet Service that provides a monolithic international service). Frame Relay is dominated by the multinational network suppliers. Each of these has built a separate network infrastructure, covering many cities in a range of countries. Unfortunately, these networks are seldom interconnected and each will offer a different range of services and tariffs. The designer will need to select a supplier that adequately covers the locations to be served, with appropriate service and an affordable price.

- ISDN—widely available and reasonably priced but can be expensive in terms of call charges. Perhaps not so well suited to LAN interconnections as either of the above two services, though, as bandwidth is more limited.

- Internet Protocol (IP)—virtual Private Networks based on Internet type technology are readily built on standard commercial offerings (routers, firewalls and the like). Each customer can be offered a private domain and other technologies can be used along with IP (e.g. ATM may underlay the IP VPN service, interconnecting its mesh of routers).

- ATM—offers the speed needed for LAN interconnect as it usually starts at 34 Mbps (although speeds down to 2 Mbps may be available). In general, it is aimed at the highest bandwidth users and can be used with SMDS, IP or Frame Relay. Tends not to be a cheap option.

There is no simple formula for choosing one approach over another—in fact, most real networks combine several of the above options as a 'mix and match' to address the varying needs.

In general, though, there are a number of drivers for selecting a particular technology—cost and performance are the two that usually dominate but customer preference and policy often have a major influence. In practice, this technology selection step is more an exercise in ruling out technologies that would not suit. For the example outlined above:

• leased lines are considered to be too expensive

• X.25 is not well suited technically

• SMDS is not well suited geographically

• ATM is too expensive and does not have enough international coverage

• IP is not yet company policy.

So, Frame Relay and ISDN seem to be the two best-suited candidates.

In practice, national tariff variations would obviously have a strong influence on the selection process. Network costs are a key parameter for most users, and tariffs vary widely over time and place. For this exercise, Frame Relay looks to be a sensible and viable option.

In verifying the suitability of Frame Relay, there are a few sanity checks that should be performed. Some of the more relevant ones that should be looked at this stage are:

Compatibility

The overall requirement in this chapter is to provide a network that connects a set of disparate and scattered resources. Hence, it is routing (where the linking of LANs is effected by switching on packet addresses), not bridging (where simpler connection of similar LAN segments is made) that is required.

Routers operate at the network layer (which is where the end-to-end network connection is set up and managed). Because they have to work with device addresses they are sensitive to the network layer LAN protocols actually used. The main extant protocols at this level in our case are IPX, TCP/IP and SNA. There are also a few small enclaves—a Mac group on Appletalk that will continue to be used in the Berlin and Edinburgh design units plus a few machines based on Digital LAT that are being phased out.

So, all of the protocols that will be required are in scope for Frame Relay, as illustrated in its compatibility table in Table 2.1.

Table 2.1 Protocols supported over a router and frame relay network

Protocol	Suitability
TCP/IP	Well understood
IPX	OK
Apple Talk	OK
SNA	Can be encapsulated with TCP/IP
LAT	Difficult (this is a protocol designed to run on a local LAN segment so is highly delay sensitive. It can only be bridged over high speed low latency links. The alternative is to use a router to translate the LAT protocol to TCP/IP Telnet protocol)

Coverage

There is a Frame Relay service available in all of the required locations. The same cannot be said, for instance, of SMDS.

Capacity

In general, the access speeds required are compatible with those offered by most Frame Relay services (e.g. in the range 64 kbps up to 2 Mbps). The company's initial intent in this case was to implement the new links shown in Figure 2.2 using 2 Mbps leased lines. The prohibitive cost of this solution triggered the revised design outlined here.

This is by no means a complete justification of Frame Relay as the most appropriate option. On balance, though, it does look like a viable approach and will be taken as the basis for implementation, subject to the subsequent more detailed checks in the rest of this section.

Now the operational details that are needed to back up the generalities thus far. For instance, can the chosen option cope with the traffic, can it go fast enough, what has to be bought, etc.? All these questions now need to be examined.

Top-level design

It is during this stage that you can start checking that you are on the right path. The user's needs (expressed in terms of services that they want) should now be shown to be viable using the technology chosen.

One of the first aspects of this is to check that the required traffic can be supported. This is a little more complicated than it sounds as there are several connotations to network traffic, the main ones that need to be considered by the designer are as follows:

- Volume—How much bandwidth is required, between which locations it is required and for how long (on average) it is needed.

- Type—The characteristics of the traffic. Different types of traffic have different tolerance to delay, as indicated in Table 2.2.

- Mix—This is the decoupling of traffic volumes and traffic types (e.g. delay sensitive, error sensitive) that is needed so that appropriate network dimensions can be planned out.

The main problem with traffic estimates is that the customer rarely knows what their requirement is. Perhaps the best way to counter this is to work from the scenarios developed earlier in this chapter and arrive at a believable set of traffic figures. For instance, by using the customer's existing traffic profiles as a baseline and then extrapolating.

Table 2.2   Broad categories of application delay tolerance

Delay seen by user	Comments
Almost instantaneous around 100 ms	This is for basic interactive work. Simple text entering on a database, operation of spreadsheet and other applications.
(note)	This means that the one-way end-to-end delay should be less than 50 ms.
1 second	This is the maximum delay for complex tasks or highly creative tasks.
1–2 seconds	A wait of 2 seconds can inhibit the user and it is difficult to remember the last task requested. In this instance, it is likely that most of the delay will be in the server or host, so the one-way end-to-end delay over the network connection should still be less than about 500 ms.
2–4 seconds	This is only acceptable if the user requests a large task to be done—a batch submission after a series of local instructions. An example here would be a file transfer. In this case, the majority of the delay is in the host to user response. The average network delay should be less than about 10 seconds.
Greater than 15 seconds	This is intolerable for a busy user, on-line. Most users will leave the terminal. In general, this is the sort of delay that is only acceptable for back-up of files at night. Batch applications and background network traffic.

This can be done by looking at the traffic level on the existing network and applying rules of thumb to come up with a required capacity for the new network. For instance, a (semi) permanent connection can be replaced with a Frame Relay link scaled to cope with peak traffic demands. If the new network is significantly different from the existing one, an alternative to this is to work out the typical number of transactions carried out by each end user and aggregate the results to give a required capacity plan. This poses more of a challenge for the designer, though, as it can be rather difficult to ensure that results line up with reasonable expectation. File transfers, remote logons and some specific operations between sites usually generate most of the traffic. Casual and random operations (e.g. mail) are usually much the smaller part of the equation.

Neither of these is ideal but they both give some sort of feel for required capacity. In our case study, we tackle this stage by sketching out the typical volumes of traffic that users are likely to generate. Part of the resulting network overview is illustrated in Figure 2.4.

Some of the traffic estimates in the figure would be fairly firmly based. For instance, it is known that the network will be used for a number of routine jobs such as sending all open workfiles to Berlin at the end of each day. These routine activities can be used, along with an average allowance for existing application usage, to put fairly accurate estimates on specific routes. But other figures, such as the traffic to and from the new site in Oslo are little more than an informed guess.

Either way, the best estimate in this case was arrived at as follows:

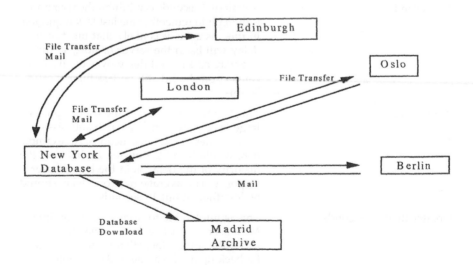

Figure 2.4 An Initial (and Partial) Picture of Traffic Flows on the New Network

- On an average day, there are 5000 file transfers from New York to London, Oslo and Edinburgh, 2000 from London, Oslo and Edinburgh to New York.

- The rest of the file transfer traffic is sporadic and fairly low level—it can be ignored.

- Files sent from one site to another are about 25 kbytes on average.

- The master database downloads from New York to Madrid at 23.00 hours each day (17.00 hours, Madrid time). This takes about an hour and is typically 50 Mbytes.

- Mail traffic between sites is fairly low level—it can be ignored.

So, the network traffic load to be supported between the main locations, at least to a first approximation is:

- From New York, 5000×25 kbytes $= 125$ Mbytes $= 1000$ Mbits. Spread evenly over a five-hour shared working day, this yields a requirement of 200 Mbits/hr, which could be catered for with a 40 kbps link to each of the three sites to which the files are sent. The likely spread in actual usage patterns suggest that the planned links should permit more like 100 kbps (this allowance would also accommodate some growth in network traffic).

- For the database download, 50 Mbytes has to be transferred within an hour to Madrid. This requires a data rate of 111 kbits/s, so this is the minimum capacity that should be built in for the New York to Madrid link.

- For the other main data transfer sites, London, Oslo (probably) and Edinburgh, the required link to New York is about 50% of the return link. About 20 kbps would suffice for each. With the same safety and growth margin as above, a link of around 50 kbps would be planned here.

- The remaining links can be dimensioned for minimal traffic.

By now, we have some confidence that the proposed network is fit for its intended purpose—at least in terms of its ability to carry traffic between all of the designated locations. The design should be fast enough to meet user requirements but not over-engineered (and therefore, overly expensive). The next step is to translate this logical plan into a physical one so that the equipment and services that make the plan a reality can be procured and installed.

Fairly early in this stage, the designer should have a reasonable feel for the traffic streams within the new network. Also the customer should be reasonably confident the proposed design will do what is required. Now we can start to put together a physical map of a network to carry that traffic, as illustrated in Figure 2.5.

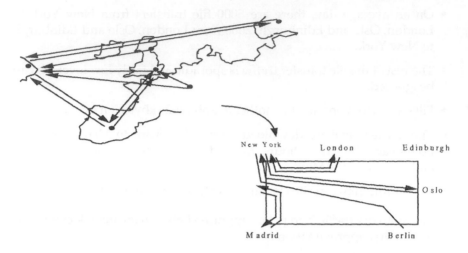

Figure 2.5   Moving Towards the Physical Network Layout

For simple networks, this can be a one-to-one mapping of traffic stream to data link. But our example (indeed, just about any real enterprise network) is a bit more complex. What we need to do now is to look at how an appropriate collection of private virtual circuits (PVCs) can be deployed to support the required aggregate traffic.

Bearing in mind that, in a typical network of this type, the majority of traffic tends to flow between branch sites and host sites (and there will be a minority flow between branch sites), there are two basic choices. These are:

- a star network centred on the host site where inter branch traffic is 'tromboned' via the host site router;

- a fully meshed network where every site can be reached directly via a PVC (which is fine initially but does not scale well).

There is also a compromise between these two—a partial mesh, where a star is supplemented by a few extra PVCs representing the more significant minority flows.

Each of these options has its pros and cons. The issues that affect the final choice are discussed in some detail over the next few chapters but the key issues that do need to be looked at are:

- Delay—for instance, there is greater delay in switching the traffic around the star network to its final destination. The meshed solution would win on this score as traffic goes directly to its destination over the relevant PVC.

- Security—it is easier to control and audit network communications when all traffic must pass through a central point. Clearly, the star network is easier to police.

- Administration—an often underestimated factor that depends on the amount of equipment that has to be looked after, the amount and location(s) of data that resides in the network, frequency of changes, etc.

- Cost—as mentioned earlier, this is more than just the cost of equipment. Services, licences, upgrades and rentals all need to be taken into consideration when costing options.

- Flexibility—how easy is it to expand or reconfigure the network. Adding another point to the star network would, for instance, be a lot easier than connecting another node into a full mesh.

In this instance, we tend towards the star option for the following reasons:

1. It is not all that likely that new sites will be added to the customer's network.
2. Administration and security are both fairly straightforward.
3. Delay is something that is deemed to be important to the customer and this option scores well here.

In Figure 2.6, each of the network sites is connected via a router to a public Frame Relay service. This provides a path from each site to any of the others. It is not the case, though, that each site has a direct link to each other. In practice, for instance, the New York site would probably be used to switch traffic around the network—the traffic stream between Berlin and Oslo would be realised as an indirect path through New York.

The arrows between sites (incomplete in the diagram, to avoid unnecessary detail) would, in reality, be used to indicate the capacity required between the relevant routers. This aspect of the design is considered in more detail

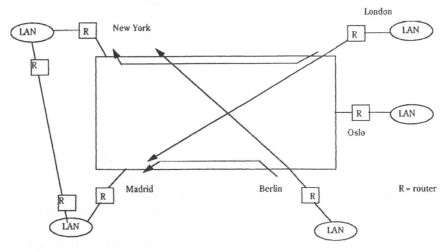

Figure 2.6 Physical Layout for the Enterprise Network

later—there are trade-offs of delay, cost etc., in choosing indirect as opposed to direct routes.

The connection between the Madrid and New York LANs is included to ensure that there is back-up access to the master database—a resilience measure that reflects the importance of the company's data. This could be realised using routers and a leased line or by having a direct PVC for the link but would most likely be a rather pricey option. An alternative option for the New York to Madrid link would be a bridge. This highlights some interesting points. With a bridge, all the traffic between these sites will of necessity flow over the bridge. Why is this? Well, bridging is a Layer 2 (MAC address based) activity. One would expect Madrid and New York to be in the same IP subnet if bridging were used. If this is the case, then routers at New York and Madrid would not be interested in forwarding any traffic from New York to Madrid—believing it to be in the local subnet. Using routers, we can use the Frame Relay path as a primary route and the leased circuit as a secondary path.

Having chosen our connection option we need to look in a little more detail at the bandwidth required between the various routers shown in the above diagram. There are two basic aspects to this—first, the speed of access link from the router to the carrier (Frame Relay) network needs to be established and, second, the end-to-end bandwidth should be determined.

With Frame Relay, there is a facility known as the Committed Information Rate (CIR). This is the bandwidth that the network supplier commits to provide over a private virtual circuit (at least, under normal network operating circumstances). It is one of the key parameters used when dimensioning links in an enterprise network. The chosen CIR, along with the speed of the access links to the Frame Relay service point provide the base for the sums that need to be done to dimension (and subsequently cost) the planned network.

There are a number of rules of thumb that can be applied in making appropriate choices here. These are as follows:

- Committed Information Rate (CIR). As a first estimate (and useful sanity check), the CIR for each PVC should be selected to match the offered data rate (although the data carried can burst over the CIR).

- Speed of each access link. At its simplest, the access circuit speed should exceed the maximum CIR (incoming or outgoing). More importantly, the sum of CIR values should not exceed the access circuit speed by more than a given factor. A set factor (e.g. 1 for delay sensitive transaction traffic like SNA and 2 for more general LAN interconnect traffic) would suffice as a design rule. This approach (known as oversubscription) may be over cautious (expensive) and an alternative would be to take a more statistical view (for example, aggregated bursty sources require less bandwidth than simple sums indicate. Forinstance, 5×2 Mbps sources, on average, requires 2×2 Mbps channels) and set the access speed to half of the greater of the CIRs into or out of network. This can be fine-tuned over a period of time with the access provider.

- Contingency. In reality, packets, like buses, tend to bunch (due to varying delay) so you should not opt for dimensions close to critical. If this is done, a little extra load can readily choke the network. The level of contingency is a design decision that varies from one case to the next. A 20% allowance is not uncommon.

Once the relevant estimates and decisions have been taken for all of the access links and PVCs included in the design, the aggregate wide area network requirements can be put together.

Table 2.3 makes a start on this. It takes each of the network sites and works out the totals when all of the incoming and outgoing links proposed in the design are put together.

This can be a fairly lengthy exercise for a network of any size. Nonetheless, by the end of this stage, the viability of the proposed design and the path to its implementation should be fairly evident. It now remains to carry out checks against the original requirements.

Implementation design

The last of the design stages in this systematic approach is to ensure that all of the above design decisions support the user needs. This is particularly important if, for instance, voice is part of network traffic. This has low tolerance to delay—50 ms for a connection with echo, 100 ms if echo control designed in—and places stringent constraints on the chosen design.

If there are problems here, the dimensioning stage needs to be revisited and, in this example, either access speed or CIR increased—or worse!

Relevant questions that need to be addressed are:

- What size buffers to use—if they are too long you get delay. If they are too short, you run the risk of losing packets.

- Is there enough bandwith to carry the required traffic?

- If all of the end-to-end delay components are summed, can you still meet the delay requirements (even the designer's estimate of what delay the users will find acceptable)?

As for tolerance to error, voice is fairly even tempered and you readily can stand bit error rates of one in ten thousand. For data, the figure depends on higher level protocols.

Table 2.3 Wide Area Network Requirements

PVC source	PVC	Traffic	Access link
New York	Madrid	111 kbit/s	
New York	Oslo	40 kbit/s	
New York	Edinburgh	40 kbit/s	
New York	London	40 kbit/s	
New York	Berlin	2 kbit/s	
	Sum of traffic	*233 kbit/s*	256 kbit/s
London	New York	40 kbit/s	
London	Edinburgh	8 kbit/s	
London	Oslo	4 kbit/s	
London	Berlin	4 kbit/s	
London	Madrid	4 kbit/s	
	Sum of traffic into node	*60 kbit/s*	64 kbit/s
Madrid	*Berlin*		
etc.			

Business Case

Although placed at the end of the process described here, this aspect
should be addressed as soon as it can be. It may not be possible to provide
precise costs early in a design, but some indications should be evident. For
instance, in this instance, there should be some feel for the Frame Relay
tariff (cost per PVC, CIR costs).

In general, the costs of network services and bandwidth vary with tariffing regions and/or the network providers. Indicative prices to back up any choices made should be manufactured!

There is actually one more step that should be included. This is that you should backtrack to the previous step in all cases of doubt, conflict or inconsistency. In any case, it is worth checking that decisions taken at later stages have not compromised those of earlier stages.

2.3   PROOF OF THE PUDDING

The ultimate test of adequacy for a network design is whether it provides the end user with what they want. Going back to the stated needs for our example, let us now see how the required services are supported. Going back to the original requirements, there are two key user functions that need to be catered for:

- the ability to update the main operational and archive databases;

- the basis for an electronic mail service covering the whole enterprise site.

To complete this first attempt at enterprise network design—the systematic approach—we consider how each of the key requirements is met by the design evolved in this section.

Database updates

Part of the working brief so far has been based on the need for a daily download of all files held on the New York database to the Madrid site. The user perception here is of a capability for baselining data by holding it on an archive machine. To be effective, this service must be reliable and resilient—if there is a network problem, the user should be able to choose an alternative route. The design evolved here can cope with this since it has a back-up route to the Madrid site.

The speed of access link designed for the New York is 256 kbps which, with a CIR to Madrid in excess of 120 kbps would provide enough capacity to complete the required transfer within one hour.

As for the rest of the database updates—those carried out during the working day, primarily from London and Edinburgh—the access circuits and CIRs worked out earlier should enable typical traffic to be supported.

There are a few more design issues that should be attended to before this application can be considered to be catered for. For instance, for each traffic stream there needs to a bi-directional PVC to allow any acknowledgements required in the transfer protocol. Also, it is worth checking that the end system applications (FTP software, database transaction processing) are in place and are interfaced to the network.

Electronic mail

The above application is fairly simple in that it is predictable in terms of frequency, volume and source/destination. Electronic mail is a little more complex in that it is an unpredictable, many to many service.

So, as well as being more complex to draw out, it is also more difficult to work out hard and fast traffic figures in advance.

Given that we do not know the traffic but do know that it is likely to be fairly light but bursty, the most reasonable approach is to do an estimate based on, say, 80 users on each LAN, each logged on for about 2 hours per day and dealing with about 50 transactions. If each transaction is around 2 kbytes, then the traffic from each node is approximately 2 kbps. Given that most mail users are not concerned by a reasonable amount of delay, we can conclude that the design gives enough capacity for the stated requirements. At least on paper, the increasing propensity of people to mail large attachments to many people may well invalidate the simple approach taken here.

For now, though, it only remains to make a value judgement as to acceptable cost for services provided (bearing in mind the need to build in flexibility), and that then is the design completed.

The approach developed in this section provides a good start for the network designer. It lays the foundations for a systematic method and has a reasonable level of inherent rigour. The only problem is that it rarely works in practice. It may look pleasingly prescriptive but practical experience is that there are a host of complicating factors that swamp pure analysis (if indeed, time is ever available for the plethora of tables and matrices that need to be produced to follow this approach in practice.

The next section introduces some of the more familiar villains that push us towards a rather different approach.

2.4 THE DIRTY DOZEN

This is not intended to be anything like a complete list of things that make network design really difficult. It does, however, represent the set most often encountered. This 'dirty dozen' is a list of the issues that bedevil (or even completely scupper) the network designers in practice. The list is compiled from some of the more bitter experiences of a number of practitioners.

1. The Klingons

What is already in place can be the curse of the enterprise network. Computers, routers, workstations, databases all cost money. Quite often they are purchased (at great cost) one year only to be superseded the next. It is not uncommon for new versions of software to render expensive routers obsolete. Often, they cannot be upgraded so have to be written off and replacements found and integrated! This, along with a natural resistance to change, means that all networks must evolve from the current situation.

The concept of a new, overlay network usually founders on the rock of data—it tends to be embedded within existing systems and needs to be coaxed out through interworking.

2. Moving goalposts

Ask someone how many steps they have on their home staircase and they will probably not know. Ask them to run up their staircase and they will do it faultlessly. So it is with networks—people can rarely articulate what they want but know the right answer when they come to use it.

It can be taken as a general rule that requirements always evolve. The initial statement of need should be seen as no more than a starting position to be influenced by:

• what technology can deliver

• what becomes clear during the design

• changes in regulation, user need, enterprise focus.

There are others but the above set highlights the need to work towards a solution, rather than leap (after much planning and analysis) from statement of requirements to delivered system. Even when requirements are quite explicit, a new service brings with it new possibilities which, in turn, modify the way in which the network is used.

3. Marketectures

The recent history of data networks has been dominated by technologies and methods that purport to allow disparate network elements to co-operate. Industry consortia and standards bodies have both gone a long way to providing 'building blocks' that can be connected together and provide user services. This does not remove the designer's obligation, though. The building blocks only really work

• at a level of abstraction slightly above reality

• if there is an overall design plan into which they fit.

Explicit choices, such as the organisation of computing resource (The Gartner Group 1987), the peer to peer communication mechanism (ANSA

1986) and the logical separation of data, processing and presentation elements all need to be made. Vendor claims for open architectures, inter-operability and conformance to standards all need to be examined carefully.

4. *Opaque trade-offs*

There are few clear cut choices in networks. There are many suppliers of the key network elements—routers, LANs, management systems etc.—and they all tend to come bundled with their own flavour of 'standard' interfaces, protocols, functions and utilities.

Even before getting to the choice of supplier(s), there are some fundamental strategic options that can be made. For instance, do the suppliers within the network have to provide open solutions (conformant to standards that are designed to allow inter0operability of equipment from different vendors) or will a proprietary choice be acceptable. The flexibility of the former may or may not be preferred to the uniformity of the latter (this choice is swayed as much by a liking for a single supplier accountable for their equipment as by any strategic development notions).

5. *Outsourcing*

Very few enterprises have complete control over all the component parts of their network. The usual case is that local equipment is owned, long haul connections leased. If the network is a global one, there are usually several network operators who contribute along with (increasingly necessary) a facilities management contractor who manages the customer and network services. All of these third parties have to be focused on what the enterprise really needs. Experience shows that this is neither easy, nor is it particularly successful in practice.

The balance between concentrating on core business by offloading (or at least sharing) the burden of running the network and retaining control is not an easy one.

6. *Underwear*

There is always some lower level technology upon which any service or any network relies. The functions of these support facilities (e.g. media access control protocols) are usually assumed to be present but are rarely accounted for. In practice, everything costs. The plethora of underwear beneath an end-to-end service can eat up considerable resource. The real throughput of routers is usually nothing like the nominal capability!

7. *Who's who*

The planning and organising of network naming and addressing strategies is a task that many people who have worked in the telecommunication industry back off from. They know just how difficult it can be. The problem is being compounded by an increasing demand among users for remote and roving access.

Yet, it is a basic user requirement to know where people are located and how they can be reached. The problem is that most enterprise networks are comprised of many disparate parts, each with its own local naming and addressing strategy. Manufacturing homogeneity from a motley (and still evolving) base is no picnic.

8. Keeping in shape

Management, maintenance and administration are vital, yet often over-looked, functions. They all require specialist equipment and people (e.g. to find LAN faults, to reconfigure user access). Administration is almost invariably underestimated (one support person per 50 users is not unreasonable—more if the applications are particularly difficult or the network very diverse).

9. Superhighwaymen

With information as *the* currency of many modern business, it is increasingly important to protect against theft or loss. The Information Superhighways bring with them the spectre of the superhighwayman and the operational threat of superroadworks. Single points of failure should be designed out, likewise, points of vulnerability. Even if you cannot avoid such weak spots, be aware of them and guard their use.

10. Icebergs

With complex networks, the devil is usually in the detail. The humble PC relies on a plethora of services between raw hardware and the user interface. There are many alternative implementations of these services and they can, potentially, all impact on that machine's interworking with a network.

The considerably less humble router is a very complex piece of network equipment. Its fairly straightforward function does little to convey just how complex it is.

Coping with this intentionally hidden, often forgotten and easily over-looked level of complexity does not come cheap. In practice, it can eat up a large proportion of the budget allocated to a network project, and this is not the only source of hidden cost. Some of the others that lurk in the detail are:

- software upgrades

- maintenance and licence costs

- administration (including, if possible, overheads imposed on users).

11. From deterministic to statistical

There is a shift of thinking required when modern, packet networks are contrasted with the more traditional circuit switched variety. The reliability of protocols, the end to end checks required, testing techniques, all need to be revisited. This is comparable to the shift in the computing world as the mainframe was superseded by client–server and distributed paradigms (Norris and Winton 1996). In both cases, you need to make sure that the same levels of control and safety are applied in this new world.

12. Technology creep

To paraphrase Elizabeth Bowden, technology does not fly like an eagle but it creeps like a rat. Some decisions and options that were once perfectly reasonable and valid become anything but that in changed circumstances. Also there is usually a non-obvious series of linkages that need to be examined before a sensible new order is restored. The way in which LANs have evolved gives a good illustration.

These days, most Ethernets are actually implemented as hub networks with twisted pairs connecting the PC to the hub. This contrasts with the original bus configuration in which each PC was connected to a coaxial cable that spanned the local net. The protocol devised for this latter configuration was CSMA/CD and it was well suited to the job as it prevented two distant PCs accessing the shared cable at the same time.

The technology creep towards hub configuration has not, in many cases, been reflected in a suitable LAN protocol and many hubs run CSMA/CD despite the fact that there are more efficient ways of dealing with access control from a central hub!

This is but one example of one piece of network technology moving without its impact being examined.

These may not all be relevant to a particular design. If none of them are issues, then the design prescription described earlier can be safely used to good effect. If, on the other hand, some (or, as is often the case, all) of the real world complications are present, then a different approach is called for. The very unpredictability inherent in some of the above make any sort of procedural design method unworkable.

The next chapter suggests an alternative to procedural rigour—the adoption of iterative design lifecycle, supported by guidelines based on practical experience. The comfort factor may be lower than the 'formula' developed through this chapter but this does no more than reflect the nature of the activity. As stated at the start of the chapter, network design is complex. To suggest that it can be tacked by set procedures would be to give false hope. Albert Einstein once said that 'Systems should be as simple as possible—but no simpler'. The same holds true for designing enterprise networks.

2.5 SUMMARY

Network design is not straightforward, not even in theory. There are a whole set of activities that have to be planned in detail, technical consider-ations that have to be analysed, proposals that need to be analysed and requirements that need to be balanced. This chapter starts by explaining the key stages that need to be included in a process that delivers a consistent and viable design. In effect, we provide a systematic process for building enterprise networks that consists of five key steps:

1. Requirements
2. Architectural design
3. Top-level design
4. Implementation design
5. Business case

This process is explained and illustrated through the development of a mythical (but not atypical) corporate network.

The key message of the chapter, though, is that this is not enough in many instances. Experience has shown that the structured approach to design only really works in retrospect (and, in any case, is only part of the overall network delivery problem). It is useful in rationalising and controlling a complex activity, but only from a distance and as part of a wider process. In short, a good sanity check but not sufficiently broad for practical application up front.

On a day to day basis it is the unexpected and unreasonable that often come to dominate proceedings, and you cannot readily codify this. The only real way to cope in the heat of battle is to combine organisation with experience. Given this, the last part of the chapter has explained some of the extraneous factors that tend to afflict the designer. How these can be countered, nullified or, at least, minimised comes next.

We now turn our attention to design through a mix of canned experience and technical facts. The next chapter reworks the network design process as less of a procedure and more of a route map. This is backed up with a raft of technical exposés, described later in the book.

Within this framework success depends on good judgement, And good judgement comes from experience, which, in turn, comes from making mistakes. If you have suffered the latter, then read no further and refer only to the particular details you need. If you prefer to benefit from the author's mistakes, take the guidance in this chapter with a pinch of salt and read on.

REFERENCES

Advanced Network Service Architecture (1986) *The ANSA Reference Handbook*. This, plus other ANSA references are most reliably accessed via the World Wide Web, URL http://www.ansa.co.uk

The Gartner Group (1987) *The Three Tier Hierarchy for Distributed Systems.*

Norris, M. and Winton, N. (1996) *Distributed Computing in the Information Age*. Addison-Wesley.

BIBLIOGRAPHY

Aboba, B. (1993) *The On-line User's Encyclopaedia*. Addison-Wesley.

Atkins, J. and Norris, M. (1998) *Total Area Networking*. John Wiley & Sons.

BCS Medallists report (1994) The global office. *Computer Bulletin*, February.

Cerf, V. (1991) Networks. *Scientific American*, September, 333–341.

Griffiths, J. (1997) *ISDN Explained*. John Wiley & Sons.

NCC (1982) *Handbook of Data Communications*. NCC Publications.

Sloman, M. (1994) *Network and Distributed Systems Management*. Addison-Wesley.

Smythe, C. (1994) *Internetworking—Designing the Right Architectures*. Addison-Wesley.

Van Duuren, J. Schoute, F. and Kastelein, P. *Telecommunications Networks and Services*. Addison-Wesley.

World Wide Web—URL http://www.analysys.co.uk The Virtual Library (This is a set of online guides covering a range of telecommunications, computing and network topics. Coverage of the Internet and related technologies is particularly strong.)

3

Introducing the Enterprise Network Lifecycle and Design Process

A mighty maze, but not without a plan.

Alexander Pope

Now that we have dispelled the expectation that you can design enterprise networks to a set formula, we return to basics. Chapters 3–10 focus on the role of the designer in the procurement and implementation of a real-life, large-scale enterprise network. It is worth reiterating that the design of these networks is a major and complex undertaking. From this point onwards, each chapter aims to provide information and build strategies for tackling a part of the overall problem in a methodical and quality manner. The intent is that the would-be designer should have gained an understanding of the practical challenges ahead, and will be armed with a methodology along with a set of checklists for the key issues.

'Murphy's Law' of network design might state that 'A useful network always fills to capacity', and the corollary of this is that 'Useful networks cannot be designed—they just evolve!' There is some truth in this! It is also true to say that user requirements, business applications and networking technologies are constantly changing—it seems at an ever-increasing rate. In the UK, as little as ten years ago, the fastest affordable packet data network connections on offer would have probably have been 64 kbps X.25. A couple of years on from this, 2 Mbps Frame Relay connections could be had, and five years ago SMDS at up to 25 Mbps was the top of the range.

ATM links at 34 Mbps and 155 Mbps are now readily available. The dynamism of technology means that custom built networks can be highly complex and expensive. It is often wise to split implementation into phases—a divide and conquer approach!

It is for these reasons that the design of an enterprise network can never be seen as a once off act, but must be viewed as a continuous cycle. It is not dissimilar to the 'Spiral Model' of software development, one of a number of well used and accepted development processes (Kerole and Freeman 1981). This is illustrated in Figure 3.1. In this, as soon as the software is delivered, if the software is found useful, the users will invariably demand fixes to any bugs and implementation of new features—so the development process becomes a cycle, leading from delivery back to requirements gathering.

3.1 THE ENTERPRISE NETWORK LIFECYCLE

For the purposes of this book, the lifecycle has been divided into a set of stages. These are clearly defined modules of work that occur during the lifecycle. The division is not absolute, and variants may be adopted by different companies and for different projects. The important issue is not the lifecycle model used, rather that one should be chosen as a framework in which to place the enterprise network design activity.

The end-to-end stages in the design, development and operations lifecycle are shown in Figure 3.2. Links between stages do not represent strict time flow, as many activities may happen in parallel. The links have more of a sense of information flow. Chapters 3–10 will describe each of the lifecycle stages in some detail—Figure 3.2 shows which chapter covers each component. In describing each stage, the main emphasis will be on the role of the network designer. Each chapter will describe methods for tackling a particular stage. The application of these methods in concert will be illustrated in Chapter 10 of this book with reference to a complex (i.e. close to real life) case study. Chapters include checklists to help the designer handle that stage, and conclude with a summary, which states the key learning points and also problem areas associated with the stage. Section 3.2 will provide more detail on the elements of this 'Design Framework'.

To help set the context in which an enterprise network designer will work, section 3.3 of this chapter contains a description of the enterprise network procurement process.

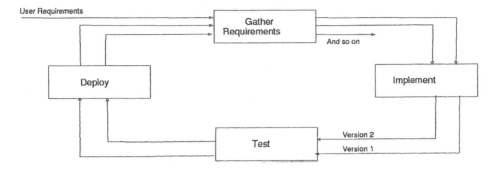

Figure 3.1 The Spiral Model of Software Development

3.2 DESIGN METHODOLOGY—DIVIDE AND CONQUER

This section outlines the design methodology that will be described in detail in Chapters 4–9. With a project on the scale of many of today's enterprise networks, the size of the design task can seem overwhelming! As with all such large tasks, a divide and conquer approach is often the way forward (Norris 1995). The design task can be divided into the following work areas (as illustrated in Figure 3.2).

- Requirements Capture and Analysis
- Architectural Design
- Physical Design
- Core Network Design
- Logical Design
- Management Design
- Verification and Validation
- Operations and Evolution

These areas of design are briefly introduced below.

Requirements capture and analysis

The first task of the designer is often to research and document the Statement of Requirements for the network. (In cases where an external

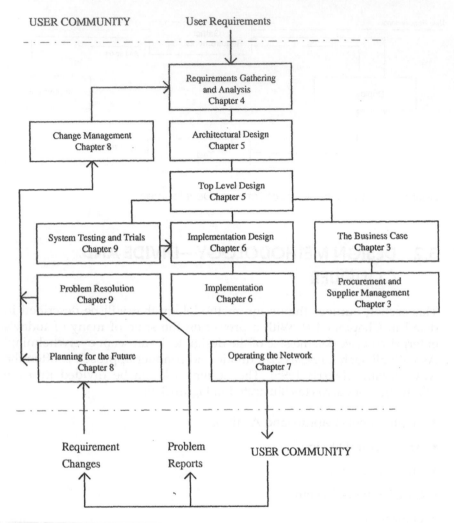

Figure 3.2 The Enterprise Network Lifecycle

supplier is bidding for a network, the Statement of Requirement is often prepared by the customer and presented to the supplier's design team.)

Statements of Requirements documents frequently run to several hundred pages, and ample time must be given over to thorough study of it. It is a good idea to use of a highlighter pen to mark up key technical requirements, as is preparing a concise written summary of the main features and constraints.

During this study, the designer is bound to discover areas where the Statement of Requirements does not provide sufficient detail. The designer should produce a written record of these queries. If it is a DIY procurement, it will be possible to go back directly to the user community to seek further detail or clarification on the requirements. If it is a formal procurement process, the customer will usually provide a formal channel for responding to written questions.

Architectural design—topology and solution component selection

Having analysed the requirements, the designer will start thinking about possible solutions in a very high level, abstract manner. Thinking at this stage will be in terms of the generic site types that need to be interconnected, rather than thinking about the in-depth detail of specific sites.

The designer will need to choose a topology, the general shape of the network, for linking the site types together in the required way. It will then be necessary to consider a range of component technologies for use in the wide area network (WAN), and for use at each site type to interface the customer end systems to the WAN. These areas will be covered in detail in Chapter 5.

It would be common at this stage to sketch out in outline a number of possible solution scenarios. A rapid assessment of the scenarios would be made in order to short list one or two possible design options to go on to the more detailed top-level design stage. In comparing the technologies, the following aspects should be considered:

- comparative costs of technologies;
- performance (network delay and throughput);
- reliability;
- potential areas of risk (especially proposed use of untested technology).

This analysis can go on to form the basis of a risk register which should be maintained throughout the project to manage the level of risk.

It is possible at this stage that we identify the need for a component in the design for which no supplier exists. This would trigger a feasibility study into the possibility of carrying out or subcontracting a custom development.

Physical design

This deals with the physical components that have to be procured and installed. This needs to be considered in some detail at the top-level design stage, as the inventory of physical components is the major capital cost for the enterprise network. Also, ongoing maintenance and management costs are generally calculated from numbers and types of equipment used in the network. This stage will need to be completed early in the bidding process so that work can start on the business case and costing of the solution.

We can divide this step into intra-site design (the local area networks and connections to legacy equipment) and the inter-site design (the wide area network). In some cases, customer end systems may connect directly to the wide area network, but in many cases, some form of network interfacing equipment will be required to connect the LAN and the WAN. In a modern enterprise network this will often be a multiprotocol router. This stage is dealt with in Chapter 5.

Core network design

The wide area network is often represented diagrammatically as a cloud—and it is all too easy to become 'cloud minded', treating the network as a thing of unlimited resource. This is not, of course, the case. Any enterprise network using a supplier's shared public network will impact on the network, and the nature of the network will impact on the design of the enterprise network. It is likely that additional network nodes and core bandwidth will be required to allow the network to cope with a large, new enterprise network using it.

It will be necessary to ensure that there are sufficient free ports in the right geographical locations. If these conditions are not met, the network supplier must plan appropriate network expansion. The supplier also needs to ensure there is adequate capacity on the various network links, and demonstrate how the utilisation of these links will be monitored, and what actions can be taken to increase bandwidth if needed. The impact of providing additional network capacity may have to be built into the bid pricing, so this aspect will have to be addressed early on. Notwithstanding, a bid would normally be based on standard network tariffs, which will have been designed to cover costs of network growth as more customers connect.

The supplier will also have to think about the optimum way of using the network. An example would be if a customer had all their traffic terminating at a single host site. It may be more appropriate to connect this host site to network nodes all around the country, rather than have multiple connections to nodes close to the host. Users would then get better performance as traffic would pass through fewer network nodes, and the network would be more resilient, as fewer users would be impacted if a node to which the host attaches fails.

Logical design

This aspect of design concerns the configuration of network components, and how these components interact as a system to provide end to end service. Key areas for consideration are design of addressing schemes, routing methods and how the network will recover under failure conditions. The principles of the logical design must be fully established at the bidding stage (to be confident that the proposed design will actually work!). The full details will generally be worked out once the business is won, prior to implementation. Details of this stage will be found in Chapter 6.

Management design

The quality of service provided by the enterprise network will only be as good as the management systems that exist to support the customer applications. Design of these systems include physical aspects (what management equipment and where will it be placed?) and logical aspects (for example, exactly how are alarms to be routed and presented, how will network components be polled to health check them, what management statistics are to be gathered?). Physical aspects of the management system will need to be settled early in the design process, so that equipment prices can be built in to the bid pricing. Management design is covered in more detail in Chapter 7.

Verification and validation

Verification consists of making calculations, tests and demonstrations to show that the design will actually meet all the customers requirements.

Validation is the formal testing of the delivered network, to ensure that it meets customer's requirements. This normally consists of an agreed programme of integration and acceptance tests done on a testbed or trial network, followed by agreed commissioning tests performed as each new site is added to the network. Verification and validation are further discussed in Chapters 5, 6 and 9.

Verification and validation will form key steps for any supplier delivering a network under a formal quality management system, such as ISO 9001. When using such a system it is necessary to be able to formally demonstrate that the product or service supplied meets the customer requirements as laid out in the Contract. Thorough and well-documented verification and validation work is necessary to meet this requirement.

Network operation and evolution

Once the network has been designed, installed and tested, it will move into an operational state. Once operational, it would be most unlikely for the network to remain static. The designer will have an ongoing role in managing change in the network. Change can come about for many reasons, including:

- Resolution of network faults (e.g. software bugs in components or design errors).

- Growth of traffic—careful traffic monitoring and capacity planning will be required.

- Changing customer requirements—for example, introduction of new applications using the network.

This post-delivery evolution can, to some extent, be catered for in the initial design. It is certainly true that failure to account for inevitable change results in a network that quickly becomes more of a liability than an asset.

3.3   THE DESIGNER AND THE ENTERPRISE NETWORK PROCUREMENT PROCESS

It may be wondered why this chapter includes a section on enterprise network procurement. The reason is that the network designer is a key player in the process—it is also an area where a graduate engineer may have had little or no experience.

As we shall see, the method of procuring the network will also have a considerable impact on who has responsibility for network design—the customer or the network supplier (Marciniak and Reifer 1991).

As with any major purchase, a great deal of planning and work goes into procurement of an enterprise network. It is not at all unusual for the procurement process to take one and sometimes as long as two years to complete, depending on the size and complexity of the network. Figure 3.3 shows a typical project plan for a network procurement. Timings will of course vary widely between projects, and this plan is intended only to set the context for the remainder of this section.

Method of procurement

Traditionally, most major corporations have designed, owned and operated their own wide area data networks. They will have purchased component

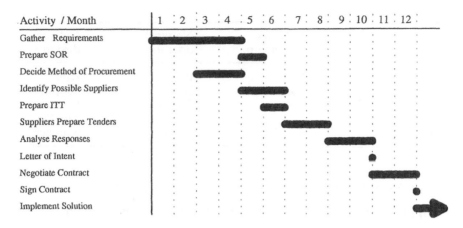

Figure 3.3 A Typical Procurement Project Plan

systems such as switches and multiplexors, and will have integrated these with private circuit bandwidth purchased from network suppliers. The company would have a team of staff dedicated to the planning, operation and maintenance of the network.

Although some companies may continue to run their networks in this DIY manner, many companies are increasingly looking to third party suppliers to provide their networking. A key reason behind this is the need for companies to remain competitive. This can be achieved by focusing on core business and driving down fixed costs, such as those associated with directly employing a networking team.

Those IT staff who remain employed by the company will be increasingly focused on supporting the core business applications. A company considering 'outsourcing' their data network will probably already have outsourced operations such as cleaning, catering and building security.

Once a company decides to move away from the concept of a network that is owned and operated in-house, a whole spectrum of possibilities open up. Some of the possibilities are summarised in Table 3.1.

Today, outsourcing is becoming an increasingly common method of procuring a network. The extent to which outsourcing is taken will vary, and depends largely upon the confidence that a company has in the outsourcing supplier to meet their business needs. A typical starting point would be that the network supplier contracts to provide a managed wide area network and managed routers at each customer site, which will interconnect the site's local area networks via the managed WAN. Possible extensions to the contract would include:

- supply and maintenance of LAN wiring and associated LAN hub equipment;

Table 3.1

	Network ownership	Network operation	Network design
DIY	User	User	User
Consultancy	User	User	Supplier
Facilities management	User	Supplier	User or supplier
Outsourcing	Supplier	Supplier	Supplier

- supply and maintenance of terminal (or client) equipment;

- supply, operation and maintenance of server computer system;

- development and operation of new applications;

- purchase and lease back of existing equipment;

- Taking over the employment contracts of company IT staff.

Identifying possible suppliers

As with most major corporate purchases, it is highly likely that the procurement will be a competitive process and that suppliers will be asked to tender (or bid) for the business. To carry out this process successfully, we first have to identify a panel of suppliers (at least three usually) who are likely to be able to deliver a suitable service, and are likely to want to tender for the business. Suppliers of outsourced enterprise networks are typically drawn from one of the following backgrounds:

- Core network suppliers—often the traditional telephone companies (PTTs) e.g. BT, AT&T or their younger rivals, such as cable TV and power line companies e.g. Energis.

- End system suppliers—usually computer system and software suppliers such as Microsoft, IBM, Fujitsu or Compaq.

- Large-scale DIY network users—these companies may oppose out-sourcing, and seek to justify their DIY network by selling it to third parties.

- Specialist outsourcing companies or consulting houses with an out-sourcing division. EDS and Andersens are typical examples here.

Likely suppliers of networks can be identified in the following ways:

- Often, if their sales force is any good, they will be self-selecting—having identified your company as one likely to purchase their services—you will probably be speaking to them already!

- Trade exhibitions and the trade press are often useful places to look for potential suppliers, especially newer companies who may not yet have approached you.

- Recommendations from industry colleagues in other companies. Membership of professional institutions can be a valuable way to 'network' with people performing similar roles in other companies.

- Larger companies and government departments may well have to formally advertise their intended procurement in an official publication, which will request interested suppliers to apply for further information.

At this stage of the procurement process, it can be desirable to narrow down the field of potential suppliers (particularly where a very large number of companies have expressed interest). A method commonly used is to write to potential suppliers outlining the type of network that we wish to procure (full requirements need not be available at this stage). The supplier is then asked to provide a statement of capability and/or give a presentation. The statement of capability might include:

- company background plus latest company report and accounts;

- details of claimed technology and network management capabilities;

- sketch design for potential networking solution;

- budgetary costs for technology elements and proposed solution;

- details of reference customers (those for whom this supplier has delivered similar solutions, and who may be approached for references).

Analysis of the resulting responses will help decide which companies are likely to be able to supply a solution and should be invited to tender.

Invitation to tender

Having arrived at a short-list of likely suppliers, a formal competitive tender or bidding process is now carried out. Before this can be done, the customer must produce an Invitation To Tender (ITT) document, to which potential suppliers must respond. A key ingredient of this document will be the Statement of Requirement (see Chapter 4 for more details on requirements gathering), which provides the supplier with the necessary information to design a solution. The ITT contains much else besides the SOR—typically it will include:

- background information on the customer, their business and IT applications;

- information on how to respond—required response format, who to deliver it to and when;

- process for querying the ITT content or raising further technical questions;

- requirements for presentations, demonstrations and trials;

- details of any specific contract clauses that may be required. An example might be where a Service Level Agreement is required (see next section).

When the supplier's network designers receive the ITT, they will have to spend time carefully reading it and highlighting the key requirements. It is likely that the designer may require further information or clarification of information in the ITT. A formal method for submitting written questions is usually used by the customer at this stage. The customer may also agree to meet with potential suppliers to answer questions and clarify requirements.

The next step for the designer will be to put together and document a top-level design (see Chapter 5). The result of this work can then be passed to the supplier's commercial team, who will cost up the design, prepare a business case and agree pricing. The supplier must have a business case, as supplying large networks normally involves a considerable up-front investment by the supplier in network infrastructure. The customer repays this investment during the contract through installation and rental charges. The business case helps prove that the supplier will make an adequate return on the investment. This return needs to be better than putting the money in a bank in order to be worthwhile!

Having carried out the design work, the network designer is now likely to be involved in preparing text for the tender response document. Usually this will be in the form of a description of the proposed network. It is common practice for an ITT to request a Conformance Statement. This requires a specific response to each clause of the Statement of Requirement, showing if the supplier can conform to the requirement or not. As well as a simple 'Yes/No' statement, text will be required to explain how the requirement is met, or, if not met, what the supplier will do about it (e.g. they may offer to develop a solution to the requirement during the first year of the contract).

At the end of the tendering period, the customer for the new network will receive tender response documents from each of the potential suppliers. The work of assessing these and selecting the 'winner' now begins. Many aspects are considered during the evaluation of the responses. Clearly, the whole life cost of each of the offers is an important factor in making the selection, but it is by no means the only one. It is also important to establish that a suitable solution is being offered from a credible supplier.

The customers technical staff will usually contribute to evaluation of the bid. They are likely to be involved in the following activities:

- studying the technical detail of each bid;

- attending supplier technical presentations and raising technical questions to clarify understanding of proposal;

- scoring the Conformance Statement. A common practice is to score each requirement (say on a scale 0 to 5 for not met to fully met). A weighting factor (say 1 to 10) may be applied to each requirement (depending on the perceived importance for the customer). Bids can then be compared quantitatively by calculating a sum of products for the score/weight pairs);

- observing any technology demonstrations or attending visits to network management facilities, supplier research and development labs etc.;

- overseeing any required technology trials.

Once the winning supplier has been selected, it is common practice to notify them of their success using a letter of intent. This is not a binding contract, but generally authorises the supplier to spend a certain sum of money on final contract negotiation work and carrying out further design work leading up to the implementation of test bed networks and perhaps pilot service at a small number of sites.

Contract and SLA negotiation

Work can now begin on contract negotiation. It is essential that the contract contains a fully agreed statement of requirements. Between the time that the Letter of Intent is issued and the contract is signed, the customer and supplier network design team will need to work to fully review and assess the customer requirements, and prepare this information into a suitable form for inclusion in the contract. If the contract includes outsourcing of an existing network, an extensive audit of that network will be carried out at this stage. Staff transferring to the new company will be interviewed and new contracts of employment will be drawn up (the law will often require these new contracts to offer equivalent terms and conditions to their existing contracts). This period between Letter of Intent and contract is often referred to as 'Due Diligence'.

Why do we need to re-examine the requirements at this stage? One reason is that many months may have passed since the original SOR was assembled. It is more than likely that there has been some change in requirements since that time. It is also likely that the SOR was assembled with only partial information available (e.g. the network may have been specified for new software applications which were at an early stage of development—more may be known now about the networking requirements of these applications). A second reason is that while the business is being tendered, a rather formal relationship exists between the customer and the supplier design team, with a relatively restricted flow of information.

It may be that, while the winning company provided the best design, the design may be less than perfect, due to supplier misunderstandings over the requirements. Now is the time for customer and supplier to work closely together to reach a common understanding of requirements and have them formally represented in the contract.

Naturally, changing requirements at this stage will usually mean rework of the bid design, and may have an impact on the price of the service to the customer.

A major part of the contract that is often negotiated and agreed at this stage is the Service Level Agreement (SLA). This focuses on the quantifiable quality aspects of the network, how these aspects will be measured, how the results will be presented to the customer, and what penalties will be payable if the SLA is breached. Typical elements in an SLA are:

- targets for time to deliver sites after they have been ordered;

- agreed site roll-out rates;

- definitions of network availability and targets;

- definition of major incidents (e.g. 10% of sites out of service for more than 15 minutes) and penalty for occurrence;

- network performance targets (e.g. throughput and cross network delay).

Given that the SLA is all about agreed performance standards, there is an implication that part of the network design should focus on measurement and management tools. The reports required by the customer go a long way to setting the requirements for these tools. Management design is dealt with in a later chapter.

Long term supplier management

Of course, network designers are not just involved during the sale of a network. They will play an active role in the relationship between customer and supplier during the implementation of the network and for the duration of the contract. Some of the areas of involvement will include:

- *Implementation planning meetings*—designers will have a key input to deciding the order in which various elements of the new network must be implemented—they will, therefore, have a significant influence over the detail of the project plans.

- *Technical meetings*—technical representatives from both the customer and supplier side can expect to have regular technical meetings. An important role of these will be to review the design as it evolves, to ensure that it will meet customer requirements. The designer also has a key role in explaining and interpreting the design to the technicians involved in its implementation.

- *Progress review meetings*—the customer will want regular progress meetings to review implementation progress against the plans. The person responsible for the design will normally attend these meetings to discuss any technical issues or network failures that are affecting the progress of the implementation plan.

- *Network testing*—the designer will work with the customer to design a set of tests needed to prove the network design, and subsequently to show that each new site added to the network is operating correctly. The designer will also be involved in analysing the results of testing and investigating any serious problems raised. Chapter 9 covers this area.

- *Contract change control*—in any long term network contract, a certain degree of change to the network is inevitable. The designer will be involved in assisting the customer to specify their requests for change, analysing the impact on the network, carrying out redesign where necessary and assisting in the implementation of the change. More detail on this aspect of the designers role can be found in Chapter 8.

- *Periodic design review and capacity planning*—the designer should be involved throughout the life of the contract in proactively investigating the performance of the network, and advising on the need to increase capacity where necessary. The designer should also explore with the customer any plans for new or changing uses for the network, so that thinking can start on the impact on the network before the customer formally requests change (see Chapter 8 for more detail).

- *Terminating the contract*—this may happen because the contract has run to term and the customer has selected an alternative supplier. In the worst case, it can happen because the network has failed to meet the customer's expectations, and the customer is using the cancellation clauses of the contract. Obviously the supplier's situation at this time will be both politically and commercially difficult. The network designer will have to maintain a professional attitude and assist in the smooth migration of the customer's applications from the current network to that provided by the new supplier. The designer will also be a major contributor at the 'post mortem' review, to help the supplier learn any lessons that can be drawn from the loss of business.

A good supplier relationship can take one of a number of guises. In some cases there is long-term commitment on both sides and contract termination would be a major event. In others, the relationship is more ephemeral with cost a major driver for change. The type of relationship should be chosen to fit circumstance (Gallliers 1991).

3.4   CHECKLIST: THE ENTERPRISE NETWORK BUSINESS CASE

Creating an enterprise network is a major corporate investment. As with any other business investment, it will be necessary to create a business case to demonstrate to the company board that the investment is a good use of company funds—i.e. will the financial benefits of the new network be significantly greater than the initial and recurring costs? Only once the board has approved the business case will funds be made available to start work on the project. Once work has started, it is then necessary to maintain the business case. This involves tracking actual spend against predicted spend, and reporting and agreeing any major discrepancies with the board. As requirements for enterprise networks evolve over time, (see change management component), the business case will need to be frequently revisited as major changes and extensions to the network are planned.

Of course, it is not just the company procuring the enterprise network that must create a business case. The supplier will also need to produce one, to ensure that the investments in capital equipment and operating costs make an acceptable financial return for them.

The exact methods used for preparing and tracing business cases will vary from company to company. It is assumed that the IT director and his staff will be supported at this stage by staff from the finance department, who will be suitably experienced in business case development and maintenance to the company's standard.

Work on the business case begins at an early stage. The customer for the enterprise network is likely to want a benchmark costing of a DIY price, even if an outsourcing approach is intended. This costing will help in judging the competitiveness of offers from outsourcing companies. For them, supplying an enterprise network to a customer will normally represent a considerable up-front investment, which must be recovered with suitable profits over the life of the contract.

This checklist contains some of the areas that the business case must cover:

Costs

❏ Capital cost of hardware components.

❏ Maintenance charges for hardware (if unknown, a rule of thumb figure must be adopted—15% of capital cost per annum is often sufficient to allow for replacement of failed components).

❏ Software licensing and maintenance charges.

❏ Staff costs (for design, project managing and carrying out the roll-out, operation and maintenance of the network).

❏ Training.

❏ Test equipment.

❏ Accommodation (floor space costs of systems and any special environmental services e.g. air conditioning).

❏ Electric power systems installation and power consumption.

❏ Racking requirements.

❏ Installation charges for building wiring.

❏ Installation, rental and usage charges for wide area network circuits.

Savings

Remember to offset costs against savings that will be made in moving away from existing network equipment. This will include:

❏ Rentals of private circuits to be ceased.

❏ Ceasing maintenance and software licence fees for existing systems.

❏ Scrap or resale value (if any) of equipment to be removed. Note that old equipment often has a residual asset value to the company (capital cost less depreciation). Equipment may be depreciated over a 3–7 year period. It is not uncommon for the residual asset value of equipment to exceed its open market scrap or resale value. This can make it difficult to dispose of unwanted equipment, as company accountants may be unwilling to write off the residual asset value.

Business benefits

As well as savings resulting from the new network, the business will also be expecting to gain business benefits, which should be costed into the business case. Quantifying these benefits can prove a difficult task! Examples are:

❑ Increased staff productivity.

❑ Increased sales/market share through new improved customer service routes to market (e.g. internet shopping) and ability to offer new products.

Note—Compiling a business case is not without difficulty. Some of the key issues associated with it are:

• Difficult to obtain accurate component costs at an early stage. This is because suppliers may not wish to discuss their final (discounted) prices until the formal bidding process starts.

• Hard to quantify value of many user benefits accurately.

• Accurate costs will not be available without a detailed design. Timescales may only permit top-level design at this stage.

• Hidden costs. All risks with the design (such as use of new equipment of software) should be assessed and the potential financial impact of each risk assessed.

3.5 SUMMARY AND KEY ISSUES

The purpose of this chapter has been to introduce the key stages of the process model for designing an enterprise network. The process requires the designer to take a logical, top-down approach to design by:

• researching and analysing the customers requirements;

• considering the overall architectural design;

• design the physical aspects of the network;

• give due consideration to the impact that the network under design will have on the proposed core WAN services. Also, consider how best to exploit the topology and services of the proposed core WAN;

• design the logical aspects of the network;

• carry out appropriate verification and validation work;

- ensure that the design task is treated as on ongoing process during the life of the network.

The chapter also sets the context for the network designer's task, by explaining the methods by which large-scale enterprise networks are procured. As such it provides a guide to the following six chapters which, together, give the designer a comprehensive set of tools and techniques to do a good job.

REFERENCES

Galliers, B. (1991) Strategic information systems planning; reality and guidelines for successful implementation. *European Journal of Information Systems*, **1**, 55–64.

Kerola, P. and Freeman, P. (1981) A comparison of life cycle models. *Proc 5th International Conference on Software Engineering*, San Diego, 1981.

Marciniak, J. and Reifer, P. (1991) *Software Acquisition Management*. John Wiley & Sons.

Norris, M. (1995) *Survival in the Software Jungle.* Artech House.

4

Requirements Gathering and Analysis

The buyer needs a hundred eyes, the vendor not one.

George Herbert

Before we can get down to the design of an enterprise network, it is essential to have a clear statement of the *real* requirements. These are not just the vague 'We need a connection in Singapore' or 'Our aim is provide ubiquitous access' type of requirement (both of these are taken from real requirements documentation). Significantly more detail should be added before engaging technical resource: it is essential to get beneath the superficial level, to really understand what applications will be used for, how they will use the network, and how they are evolving. It is equally important to understand fully what systems the company already has in place.

Final responsibility for production of the Statement of Requirements lies with the staff of the company IT department. Having said that, the gathering of requirements is a tricky business that benefits from a systematic approach and consideration of a variety of views. It is not uncommon at this stage to solicit assistance from networking companies (one with which the company has a good relationship, and is expected to bid for the solution at a later stage). Alternatively, external consultants may be hired for the task. Either way, the job of gathering requirements should not be taken lightly.

The mission of an enterprise network is to support business-critical applications in an efficient and cost-effective manner. In preparing the Statement of Requirements for the enterprise network, it is essential that the applications to be supported take a leading role. Each application should be studied to determine where it will be used and what end systems are used to support it. Detailed study of the application and its end systems will lead to the requirements of the network, such as:

- end system locations and quantities
- end system interface types and protocols
- traffic profiles, throughput and response times
- availability requirements.

This chapter explains how enterprise network requirements can be systematically gathered and applied. Our overriding requirement in writing the chapter was to illustrate the main sources and types of information and how they can be analysed and assembled so that the designer has something firm to work from.

There is always the temptation to foreshorten (or even forget) the requirements gathering process. Experience shows that every missed, assumed or misunderstood requirement costs ten times more to fix later than to get right up front (Norris and Rigby 1992). Time spent in this phase pays dividends!

4.1   STRATEGIC REQUIREMENTS

When gathering the requirements, it will be found that there are three main types: strategic, tactical and factual. The strategic requirements are most likely to be expressed by senior company management (e.g. board members such as the IT director)—the people with the 'Vision Thing'. These requirements are likely to be longer term and broadly stated. There are a variety of types of strategic requirement. The main ones are listed below, along with a few of the key issues that should be addressed.

Applications strategy

If a company can standardise on the choice of software that is deployed on the end user (client) PCs throughout the organisation, then interworking is facilitated and support costs can be minimised. A typical strategy might be to use Microsoft Windows NT on the desktop PCs. and Windows 98 on laptops. The company may standardise on a particular suite of office automation tools (e.g. Lotus Smart Suite) and a Web browser (e.g. Netscape Navigator).

As well as the basic office automation tools, users will need to access specific business applications. What are the strategic applications (existing or to be introduced) for this company? What is the relative business importance of each? The answer to this question will be a good guide to prioritising the migration of applications onto a new network.

We should expect (or, at least, hope) to see a strategy in place for the development of new applications. For instance, the enterprise resource management package SAP R3 may be the product selected for implementing business applications, or the Oracle SQL server may have been selected for database applications. Many new applications may be Web-based, so the company may have standardised on the Microsoft IIS web server software.

Protocol strategy

While modern networks are capable of multiprotocol operation, this is not achieved without cost. The equipment that is needed for this will be more expensive (needing more powerful processors and more sophisticated software). It can require considerable effort from support staff to configure the network in such a way as to balance the conflicting demands of the different protocols and compromise is inevitable (Held 1996).

Use of a single core network protocol (most likely to be TCP/IP) can lead to the most efficient and cost-effective networks. We must establish if the company has a protocol strategy in place. If so, are there other existing related corporate networking standards, e.g. naming and addressing conventions, which the designer should know about and to which they must adhere?

Open v. proprietary

When selecting hardware, software and network protocols, there will be a choice between proprietary and open systems approach.

Proprietary systems are those originally developed and supplied by a single manufacturer. A typical example is IBM's SNA. In proprietary systems, interface specifications between components, and functional specifications of these components are not always available in the public domain. Therefore there will be limited support for these components and their interfaces by third-party equipment suppliers. With proprietary solutions, the tendency is for a company to become 'locked in' to a specific supplier. This is counter-balanced by the near certainty that all components will interwork—and there is only one supplier to blame if they do not!

Increasingly, users are turning toward open systems, where standards for components and their interfaces are freely available. Opens systems offer a much wider range of supplier, with the promise of inter-operability between components from different suppliers. This is a claim that should be tested in the lab before committing to a large multi-vendor solution (Norris and Winton 1996)!

Hardware strategy

The network designer will need a good understanding of the company end-system hardware strategy. This is because we have to understand what physical interfaces (e.g. LAN or serial ports) these systems have; the design of the enterprise network must allow for connection to these interfaces.   The selection of hardware may also determine what network protocols the enterprise network must support (e.g. use of Digital computers suggests the need to support DECNet and use of Apple computers suggests AppleTalk).

The information technology marketplace offers its customers a very wide variety of end-system hardware solutions. The selection of new hardware is often determined by the application software that it will support. The company many already have a heavy investment in existing hardware—so-called 'legacy hardware' (also known as cherished hardware). If this is the case, it will often be important to continue using this hardware and this can obviously impose constraints on the adoption of new application software.

In large companies, there is always the risk that individuals or departments will go their own way in selecting end system hardware. This should be avoided, as it inevitably will increase support costs. It will increase the range of physical interfaces and protocols that the enterprise network must support. It will also lead to incompatibilities, preventing users accessing some applications on the network.

Ideally a company will lay down a firm strategy for the selection of hardware. For example, It may specify Intel processor-based PCs for end-user desktops (often a minimum build specification will be given such as Pentium II processor with 64 Mbytes of RAM and at least 2 Gbytes of hard disk space). DEC Alpha servers may be a preferred solution for file, mail systems and database servers, while Sun systems are used for Web servers. Large companies may specify IBM mainframes to run key business applications.

To pioneer or follow trends

Companies can be divided into value seekers or economy seekers (Atkins and Norris 1998). Value seekers will prefer to work at the 'leading, bleeding edge' of technology. They will frequently be driving suppliers to participate in beta trials of the latest products and services. These companies hope to gain and maintain a competitive advantage in their field of business by using the latest and best technology. The downside is that they have to be prepared to accept the technical risks involved in new technologies and services. They will also have to accept higher support costs.

The majority of companies, the economy seekers, are likely to prefer a more cautious approach, buying in to new technologies when they are at least a year old. By this stage, the suppliers will be able to provide customers with reference sites, as proof that the technology works. Technical risk and support costs are much reduced. Any competitive advantage in using this technology is likely to be diminished by this stage.

DIY v. outsource

One decision that will shape the entire enterprise network procurement process is the degree to which the network is to be outsourced (supplied, operated and maintained by a third-party supplier). This is not a simple issue as there are several 'flavours' of outsourcing (e.g. from the buying in of all network services to facilities management, where the third party administers the network on the owner's behalf).

Each arrangement has its own dynamics, risks and opportunities (Marciniak and Reifer 1991) and these need to be appreciated by the designer, as they have significant impact on the way in which the network will run.

Management systems strategy

Customers for outsourced data networks expect their network supplier to manage both the core network and on site routers using state-of-the-art management platforms. Customers also like to have a view of the network in their own operations centres, so that when faults are reported, the customer IT staff are able to determine if the problem is in the network or with the end systems.

What kinds of management systems and data feeds is the customer expecting the network supplier to provide? What sorts of reports are required and what sort of service level agreement needs to be supported?

Network administration

In a large, business-critical enterprise network, it is important to have clearly defined principles and owners for the various tasks within network administration. This is one aspect of enterprise networking that is frequently overlooked and usually under resourced.

Proper network administration is absolutely *critical* to the success of the network. Some areas that are key are addressing and naming administration, change management and security management. At this stage we need to understand who in the customer (or third party) organisation has the responsibility for these areas and how they discharge their role.

Security strategy

Companies should all have a clearly defined security policy. With tough new data protection legislation being introduced in the UK and other countries, the need for a sound security policy has never been greater. This will define how company information should be handled and what the implications are for networking applications. The network designer will need to know what requirements must be met before applications can be networked. For example, a company may be networking commercially sensitive data. They may insist that this data be encrypted before passing over a wide area network between sites. The company may also require the network design to undergo a formal security audit before it can be used.

The increasing use of the Internet by companies, to share information and carry out business with individuals and other companies, has resulted in a significant increase in security threats. We would expect a company to have a clearly documented policy concerning exactly who is allowed access to which systems, and precisely what they are allowed to do once they have accessed the system. The network designer will have to use solutions such as firewalls and access control systems to implement this policy (Cheswick and Bellovin 1994).

In the UK, BS7799 is the main standard used by commercial companies as a basis of their security strategies, and many companies are expected to seek certification of their systems and networks against this standard. Similar security standards will exist or are likely to be introduced in other countries.

4.2 TACTICAL AND FACTUAL REQUIREMENTS

Tactical requirements are more likely to be found when talking to the user departments—the people at the sharp end. These are likely to be short term in nature, focused on existing problems that the users hope an enterprise network might fix. Examples might be:

- transferring a file from A to B takes too long;
- the path from C to D is too unreliable;
- we have e-mail in this department but cannot talk to the HQ e-mail system, as it is incompatible;
- the legacy systems cannot be immediately replaced and must be supported by the new network.

Factual requirements are hard, quantitative technical facts. Some examples might be:

- lists of sites and their addresses to be supported;
- documentation of existing network;
- traffic statistics from current network;
- projections of traffic levels from new applications;
- required availability information for various networked applications;
- throughput and delay requirements for applications.

Gaining an in-depth understanding of the company's existing enterprise network will help the designer to understand the legacy applications and systems that need to be supported, as well as the traffic patterns that these generate. This is no simple task as documentation of existing systems and sites may be poor and a considerable amount of work is sometimes needed to improve this. The effort is usually worthwhile, though, as it can reveal the root causes of problems that users experience today, such as poor performance or reliability issues.

To illustrate the point (and the level of effort needed to assess the here and now), we need to have in mind that users and application designers often have little idea of protocols in current use. Hence, they are not usually aware of the levels of traffic generated by applications on the networks. For instance, programmers generally invoke communications via very high level Application Programming Interfaces (APIs), and often have little understanding of the protocols invoked by these calls or the level of traffic produced by them.

Connecting a LAN analyser to existing LAN segments can be quite revealing. Many applications are designed assuming they operate over a local LAN segment running at typically 10 Mbps. With this relatively large capacity on tap, the applications can be quite inefficient in their use of the bandwidth, sending many messages between client and server to achieve the simplest of tasks. When client and server are separated by a wide area network link running at a rather more modest 64 kbps applications often perform poorly! Some benchmarking measurements of existing networks may be necessary to establish quantitative information such as message sizes, traffic volumes and currently achieved transaction response times—this being relatively easy for legacy systems.

Of course, in some cases, there will not be an existing network. This could be the case for recently formed companies or those that are 'late adopters' of networked information technology. In many ways, it is much easier for the network designer to create a solution where there is no existing network, as there are no legacy network integration or traffic migration issues to consider.

On the downside, a company that has not previously used networking can be expected to have much less accurate views on the traffic flows that they will generate once the new network becomes operational. In this instance, the use of prototypes and the building of usage scenarios become more important (Jacobson *et al.*. 1993). Anything that introduces a direct, hands-on feel to the network is useful.

4.3 METHODS FOR REQUIREMENTS GATHERING

Gathering requirements for any complex technical artefact is not easy. As we have already hinted, there is usually something of a gulf between the initial, fairly vague statements that are discussed when a contract or tender is being let and the detailed and consistent definitions that the designer needs to work from. Bridging the gulf calls for both technical and organisational understanding. The former, rather obviously, is needed to enable the art of the possible. The latter ensures that any technology that is delivered will actually fit. A number of requirements capture and analysis methods have been developed over the year, most of them to help with software development projects (Norris and Rigby 1992). There are a few common themes that run through all of the methods:

- Taking the systems view—linking technical requirements to the business environment in which they exist. This can avoid the technical solution looking for a problem and helps the designer make appropriate choices.

- Engaging all stakeholders—identifying all of the people affected by the technology and treating them as a part of the overall system, with valid inputs and desired outputs. In short, making sure that each stakeholder has their say.

- Enforcing change control—simply keeping track of requirements, recording changes to them along with the reasons for change. As well as maintaining an audit trail, a systematic approach helps in optimising the final product.

Getting a set of requirements for an enterprise network can be a politically sensitive business and so benefits from the above principles. The main driver behind this is that the introduction of a new network will (most likely) mean significant change and unless handled with care, people can be very upset by the prospect of change. As Nicolo Machiavelli observed some 500 years ago, the benefits of change lie in the future and are yet to accrue whereas the costs (along with the threats) are very much in the here and now.

In the case of a new enterprise network (with its attendant applications) being introduced, the likely changes include:

- Different working methods—people will need to be retrained and, in some cases, redeployed.

- Possible staff cuts or the transfer of employees to a different employer—especially if plans include outsourcing.

- Undermining of existing powerbases—in particular, a company-wide information system can remove reliance on a particular function or department.

- A change of operational dynamics—when the speed or capability of networked applications make it possible to do things faster or better.

It is therefore important that the requirements gathering process is carried out in a sensitive manner, with the full support of departmental managers. Every opportunity should be taken to present users with the benefits of the proposed enterprise network, informing and winning over staff where possible. Lack of early information can be one of the main causes of resistance to any change at all (let alone the one that the designer is planning)!

Requirements gathering can be effectively carried out through a series of interviews, with senior managers, departmental user representatives and specialist users or technical staff. Such meetings not only provide a chance to gather the required data, but also give the chance for departmental staff to find out something about the changes to come, and feel more involved in the process.

It can be useful to send those to be interviewed a questionnaire in advance of the interview. This will prompt the interviewee to gather any necessary information beforehand. Sending out of the questionnaires should be agreed by senior management, and preferably be accompanied by a letter from them, explaining the reason for this extra workload, and the importance of it to the company.

The questionnaire should be kept as simple as possible, so as not to intimidate the recipient. It should serve to prepare them for and act as a basis for the interview sessions. Interviews with relevant staff can then be scheduled, allowing sufficient time for the required information to be gathered. It is recommended that after each interview a report be written, while the information gathered is fresh in the mind.

As mentioned earlier, a key source of information for the design of an enterprise network is details of the existing network and the systems attached to it. There will often be strong financial pressures to support these old systems over the infrastructure provided by the enterprise network. For such systems, often referred to as 'Legacy Systems', it is important to gather and audit the completeness of information on what currently exists at this stage.

4.4   REQUIREMENTS ANALYSIS

Once all available information has been gathered, work can commence on analysing the requirements as a precursor to the designing of the new enterprise network. When analysing the requirements that we have gathered, we often (in truth, invariably) encounter gaps in the information available or in our understanding of it. Any such gaps should be recorded, and referred back to an appropriate representative of the company for which the network is being designed. If this is not possible (perhaps due to the constraints of a formal tendering process), or the customer does not know the answer to the question, an assumption must be made to cover the gap in knowledge. It is important that any assumptions made when designing an enterprise network are explicit, clearly documented and reviewed with the customer.

A key deliverable from the analysis of the information gathered will be the production of a formal Statement of Requirements document. This consolidates all of the gathered information into a consolidated and formally documented set of requirements, which the customer can review and sign up to. An overview of the information that should be covered in this document is presented in the checklist section later in this chapter.

As well as the in-depth presentation of requirements, the Statement of Requirement should be prefaced by a management summary, presenting the key findings and recommendations. It is likely that the person responsible for gathering the requirements will have to give a board level presentation at this stage, which can act as a high-level review and sign off for the requirements gathering stage.

During the requirements gathering phase, large amounts of material will be gathered. Organising this into a Statement of Requirements can seem quite a challenge—and it is! The requirements gathered will have to be analysed and organised into a logical and understandable structure. Presented below are some methods that will be found useful in thinking through and organising the requirements.

Analysis of strategic requirements

The strategic requirements should give the network designer a sound understanding of the medium to long-term requirements for the network. At this stage we should understand:

- The customer's plans for developing new applications—and hopefully a high-level understanding of how these will work—especially in terms of how traffic will flow between sites as a result of the new applications.

- What end-system hardware and software platforms will be used to support these applications—and as a consequence, the set of network interfaces and protocols that must be supported.

- We will understand what level of risk the customer is prepared to take—they may relish the challenge of using leading edge technology to gain competitive advantage, or they may prefer a safer approach using tried and trusted solutions.

- We will understand how the customer intends to procure the network—perhaps a DIY solution or maybe a fully managed outsourcing deal.

From these requirements, we will need to develop and document an overall architecture to describe and shape the proposed enterprise network. This will present a high-level view of how sites should be connected together and what technologies will be used to achieve this. More details on developing the network architecture are given in Chapter 5.

Site type analysis

A method that can help to organise information at a more detailed level is site type analysis. A typical large enterprise network could have from several hundred to several thousand sites. During the early stages of design, it is not possible to consider each site in detail. It is more appropriate to think in terms of a minimum set of generic site types. Indeed, best economies of scale will be achieved with the design if we can stick with standardised site designs wherever possible.

For a bank, we might have the following site types:

- Remote cash point (ATM) sites
- Bank branches (perhaps divided into small, medium and large)
- Regional offices
- Headquarters
- Computer centres
- International dealing offices.

For each generic site type, the requirements should clearly identify:

- How many sites of this type are there?
- What applications the sites run.
- What legacy systems need to be supported.
- What new systems need to be supported.
- What levels of service availability will be required at that site.

An outline design for the more important site types along with an overview of how applications and services operate across the network provides useful early sight of the new network. Simple animation and simulation at this stage can help clarify requirements, especially in the instance where there is no existing network against which to gauge requirements.

Traffic flow analysis

Any network will have to be designed to support the worst-case traffic flows. We can use the concept of site types to draw up peak traffic flow diagrams or matrices for each key application. These can then be merged to identify the worst case traffic flows required between each of the site types—using imagined figures for our banking example Table 4.1 shows how this might be. In this table, the figures in the cells are intended to represent peak traffic flow requirements in bits per second. The figure represents the typical flow of traffic from site types shown in the 'From' row toward the site type shown in the 'To' column.

Further calculations will be needed to identify bandwidth required at a destination site. It will be necessary to sum the products of all source site traffic flows with the number of that site type delivering traffic to that site.

Table 4.1

From/to	Remote ATM	Bank Branch	Regional Office	Headquarters	Computer Centre	Internat. Dealing
Remote ATM	0	100	0	0	0	0
Bank Branch	500	0	25k	0	20k	0
Regional Office	0	35k	0	10k	120k	0
Headquarters	0	0	20k	0	120k	50k
Computer Centre	0	10k	120k	120k	2M	20k
Internat. Dealing	0	0	0	100k	20k	10k

At each site, we need to consider the total traffic flowing from the site and the total traffic arriving at the site. The higher of these two figures will determine the bandwidth requirement for the site. Chapter 5 will describe how best to accommodate this peak demand in terms of wide area network access circuit speeds.

This is a rather simplistic approach to determining peak bandwidth needed at a given site, but is often the best that can be done with the level of information typically available at this stage. Ideally, we will have detailed time of day, day of week and week of year traffic profiles for each application—if this is the case, we can add the traffic profiles together, finding the peak demand when time of day is taken into account.

For example, batch file transfer applications may run at night and interactive transaction traffic mainly during the day. E-mail runs during working hours, with peaks first thing in the morning and after lunch. A security application may generate constant background traffic throughout the day—see Figure 4.1.

This graph shows the combined application traffic demands. The peak is in the 1400 to 1500 time period, at which time 44 kbps are required. Simple addition of individual peaks, ignoring time of day, would have suggested a requirement of 84 kbps! Note that this graph shows traffic arriving at a branch site—a similar analysis would be needed for traffic leaving the site for each application, and the greater peak used in subsequent design (access circuits into wide area networks are generally 'full duplex' with equal bandwidth provided in either direction).

Although it is not the case in this example, the author's experience suggests that bandwidth requirements are often dominated by batch transfer requirements. Banks often need to upload large transaction logs from branches each night. Retailers have to upload sales information from their EPoS (Electronic Point of Sale) terminals, and download them with

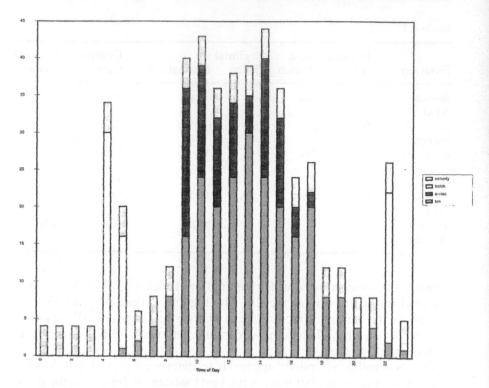

Figure 4.1 Peak Traffic Analysis for Traffic Arriving at a Branch Site

updated price look-up tables. A lottery operator may be most concerned about being able to download all terminals with new software quickly, to launch new games or facilities. The windows available for these batch file transfers are often small—data collected from bank branches and shops will need to be processed overnight with results ready for next day's business. It is often found that when sufficient bandwidth has been provided for these transfers, there will be more than adequate bandwidth available for the real time transaction traffic which tends to dominate working hours. It is therefore wise for the network designer to probe fully the requirements for these batch transfers.

Service requirements analysis

The service requirements of the network (e.g. network availability, repair time, change management and so forth) are at least as important as the technical requirements considered above. The levels of service are likely to vary according to application, location and organisation—an intercontinental funds transfer is probably significantly less tolerant of delay or downtime than a departmental newsgroup.

A useful way to analyse these requirements for the SOR would be to start thinking in terms of the kind of Contract and Service Level Agreement that the customer will want to have in place. The requirements can then be organised into the form of clauses in a proposed draft contract. As we will see later, higher levels of service tend to cost more (for example, greater reliability may entail back-up circuits). Sometimes this is worthwhile, sometimes not—it is well worth bringing such issues to the fore at this early stage.

4.5   CHECKLIST: WHAT GOES IN THE STATEMENT OF REQUIREMENTS?

This checklist is intended to serve as an aid when gathering customer's requirements—it is built on all of the advice given earlier in the chapter. The approach we have taken is to first understand the customer's business and the applications they wish to support on the new network. We then seek to understand the end systems involved in supporting those applications. Next we look at the customer's sites, to understand where they are, how they can be divided down into generic site types and what applications and end systems are present at each site type.

We conclude by looking at the existing network and any specific requirements that the customer has for the new network.

The customer

❑ Obtain an overview of the customer's business.

❑ Obtain an understanding of the business sector in which the customer works—for example, who is the competition, what are the threats and what are the opportunities.

For each application (current and planned):

❑ Has a description of the application been provided?

❑ Has the business impact been assessed? (Is the application 'mission critical'? What is the impact of failure? What priority does the application have for integration into the enterprise network?)

❑ What end systems (e.g. hosts, servers, workstations) are used to implement each application?

❑ Has the applications mode of operation been described? (Types of operational mode are real-time transactions, batch transactions, file transfer and client–server, which are similar to real-time transaction applications but may involve substantial data transfers in response from the server.)

❑ Are typical message flows for the application understood, including sizing of typical messages and files to be transferred?

❑ Are traffic profiles understood? (This will include time of day, day of week and week of year variation where known. The essential information is to establish the timing and magnitude of the traffic peak for the application. For any dial-in traffic the important factors are calling rate, average holding times, required blocking rate).

❑ Has any expected growth in this traffic been described?

❑ Have user requirements for transaction response time, file transfer throughput and any other performance criteria been documented?

❑ What is the priority of this applications traffic over that of other applications? Are there concerns about interaction between this applications traffic and that of other applications? (An example here is that of a network access circuit carrying transaction-oriented traffic and file transfer traffic simultaneously. If priorities are not correctly set, the performance of transaction-oriented traffic could become unacceptable during the file transfer. Even worse, if the transaction-oriented application is session oriented, the session could time out, forcing the user to have to log on to the application again!)

❑ What are the availability requirements? Do these vary with time of day and day of week?

❑ What application and end-system strategies exist to handle failure (e.g. is dial back up required, or can applications revert to local working?)

❑ How exactly is availability to be defined and measured?

For each application end system:

❑ What is the make and model?

❑ What types and speeds of physical interface does it support?

❑ What communications protocols does it support? (Be specific about protocol version and the values of any protocol parameter settings and facilities that are to be supported.)

❏ Are any special protocol conversion facilities required in the network? (For example, the network may need to attach an asynchronous, serially connected character mode terminal to a LAN connected host running TCP/IP protocols. A terminal server providing Telnet support will be required for this conversion.)

Customer sites

❏ Obtain a full listing of customer sites requiring connection to the enterprise network. (Note that WAN suppliers will need zip code (post code in the UK) and contact phone number for each site, as these can be used to locate the sites on a map and work out distance-dependent aspects of circuit costs).

❏ Can the sites be categorised into a set of generic site types (each site within a specific site type grouping will have similar requirements to other sites within that grouping)?

❏ For each site type, what applications are used?

❏ For each site type, what end systems must be supported?

❏ Estimate degree of site churn (e.g. what percentages of sites in any year change, to opening new sites or closing old ones). Churn can be a significant cost factor in the retail sector. It can also become high for strategic reasons—for example, the company may have to relocate all headquarters staff out of the country's capital city during the life of the project, to reduce accommodation costs.

Strategic requirements

❏ Is the customer following open or proprietary standards?

❏ What applications strategy is in place? What are the important applications to the business and their relative priority? What standards are being followed in the development and deployment of applications?

❏ Is there a strategy for networking protocols? Are there any related corporate networking standards (e.g. naming and addressing schemes) which the designer must follow?

❏ What hardware standards are being followed?

❏ Is this customer inclined to technology pioneering, or do they prefer to follow established trends—are they value or economy seekers? Either way, it is not uncommon for a customer to indicate a strong preference for a networking technology or component vendor to be used as part of the enterprise network. The network designer will need to be aware of and sensitive to these prejudices.

❑ Are they seeking a DIY or outsourced solution (or some midway point between these two extremes)?

Enterprise network management requirements

❑ What management systems does the customer require to provide real-time reporting of network status?

❑ What statistics must be provided to assist with network planning, and to report performance and availability back to users.

❑ What tools are required to configure, document and manage change in the enterprise network?

For existing network:

❑ Has the topology of the existing network been adequately documented?

❑ For each existing application, has it been shown how this is supported on the current network?

❑ Have any migration issues been addressed (e.g. need to upgrade end systems and the need to minimise system downtime)?

For the proposed enterprise network:

❑ Have any current proposals for the topology of the enterprise network been documented?

❑ Have any approaches to the design of the enterprise network been rejected? If so, why?

A good understanding of all of these points provides a firm basis for technical design. In reality, there will be issues that are not clear or information that is not available. To cover these gaps, a set of explicit assumptions should be made and agreed with the customer. Once that is done, design can proceed with a modicum of confidence that effort will be well spent.

4.6 SUMMARY

The temptation to dive into technical detail bedevils many a high-tech project (and is often cited as a prime cause of failure). To be effective, the network designer should have a detailed Statement of Requirement to work from. In many cases, network design staff will be directly involved in producing this.

This chapter has shown how one might go about producing a good Statement of Requirement—one that has enough detail for a designer from which to make sensible proposals. This chapter has also detailed the main areas that must be covered in the Statement along with a variety of techniques for making sure that they are adequately catered for.

REFERENCES

Atkins, J. and Norris, M. (1998) *Total Area Networking*. John Wiley & Sons.

Cheswick, W. and Bellovin, S. (1994) *Firewalls and Internet Security: Repelling the Wiley Hacker*. Addison-Wesley.

Held, G. (1996) *Understanding Data Communications*. SAMS Publishing.

Jacobson, I., Christerson, M., Jonsson, P. and Overgaard, G. (1993) *Object Oriented Software Engineering: A Case Driven Approach*. Addison-Wesley Longman.

Marciniak, J. and Reifer, P. (1991) *Software Acquisition Management*. John Wiley & Sons.

Norris, M. and Rigby, P. (1992) *Software Engineering Explained*. John Wiley & Sons.

Norris, M. and Winton, N. (1996) *Energize the Network*. Addison-Wesley Longman.

5

Architectural And Physical Design

*All government, indeed every human benefit and enjoyment, every
virtue, and every prudent act is founded upon compromise and barter.*

Edmund Burke

Many of the decisions taken early on in a project will have significant
impact on the look, feel and quality of the final product. The designer has a
very important role to play at this point. They must ensure that the
transition from outline requirements to a real, operational network is
soundly based. This means making sure that high-level designs are
achievable, practical and appropriate. This is not a simple job and calls for
a mix of experience, analysis and know-how.

Fortunately, there is a wealth of experience in network design that we
can call on. There is the slight problem that most of it tends to reside in
the heads of much sought after individuals but this chapter offers a
solution. We will build up a systematic approach for top-level design
over the next few chapters. In doing this, we will explain the key issues
that need to be addressed, the options that are available at each stage
along with the pros and cons of each one and the sanity checks that can
be applied.

5.1   INTRODUCTION AND OVERVIEW OF ENTERPRISE NETWORK ARCHITECTURE

The objective for network design during the architectural phase is to establish and document a top-level view of the proposed enterprise network, with sufficient level of detail to:

• Establish and document the architectural approach to be used in the solution.

• Demonstrate that the design will meet user requirements.

• Enable the design to be costed with reasonable accuracy.

• Identify major issues that need investigation and resolution before implementation.

The prime emphasis for the moment will be on physical aspects of the design, and methods for physical design are covered in depth in this section. Other aspects of design (the logical, management and core design) are covered in later sections.

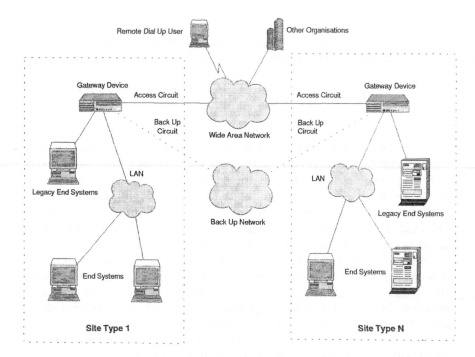

Figure 5.1   An Abstract Enterprise Network

At this stage in the design process, the designer will be thinking in fairly abstract terms about the network. Figure 5.1 shows such an abstract view of a typical enterprise network. The network has the following features:

- Generic site types. The concept of generic site types was introduced in Chapter 4. In this case we have shown two, site type 1 and site type N. An efficient network design will aim to have the minimum possible number of site types. A solution can then be designed for each site type, and that solution rolled out to all instances of that site type. Site type 1 in this diagram could be a bank branch and type N a host site.
- End systems. Ultimately, it is the applications running on these that the enterprise network must interconnect. End systems are often modern devices, which are normally interconnected via local area networks (LANs)—typically clients (workstations) and servers based on PC technology. However, the enterprise network may also have to support devices of older technology—Legacy End Systems—and these are often interconnected using serial interfaces. Examples of this would be IBM 3270 screen mode terminals and their associated cluster controllers.
- Remote users. Today, there are increasing trends toward a flexible workforce   employees may be based at home or spending increasing amounts of time away from their offices, at their clients' locations or travelling. It is becoming increasingly important for these flexible workers to have access to the same computer applications as their office-bound colleagues. The enterprise network designer is faced with accommodating an increasing number of dial up remote working users.
- Other companies. Today's businesses are increasingly trading with one another electronically—often referred to as E-commerce. This can only happen if companies are able to exchange data between their respective networks. The facilities for doing this are often described as an extranet or a Community of Interest Network (COIN).

The type and number of the above features are usually fixed in the Statement of Requirements. The designer's task is to develop a network design to interconnect them in such a way as to deliver the required end-to-end applications. To achieve this, there are a number of piece parts available to the designer:

- Local area network. This is effectively the on-site wiring system, used to interconnect the modern end systems on the site to each other. The designer will be interested in selecting topologies and technologies appropriate to each site type.

- A wide area network. The WAN is required to interconnect the sites. At this stage, the designer will be looking to select a topology and technology for this network.

- A back up network. This is an optional feature, and would be required in mission-critical networks where the main wide area network is unable to provide the required level of availability or repair times.

- Access circuits and back up circuits. These connect the sites into the wide area network, and as such, are normally an integral part of the WAN service provided by a network supplier. The designer will be interested in selecting the optimal transmission speeds for these circuits.

- Wide area gateway device. This device is used to adapt the protocols and interfaces used on site to those used in the wide area. Again, the designer will need to select an appropriate technology for this device, depending on the choice of local and wide area networks, and the communications protocols employed by the end systems. The gateway device is often used to service the legacy end systems as well as the modern LAN attached devices.

Each of these needs to be considered in the design as they will all play some part in the deployed enterprise network. In this chapter we will structure a systematic approach to designing the architectural and physical aspects of the enterprise network. The approach is presented in the following sections:

- LAN design for small sites. This section will overview LAN architectures and the common types of LAN that the designer is likely to have to work with. It will show how LANs are typically designed and installed on small sites.

- LAN design for large sites and campuses. This section will look at how LANs can be scaled for size and performance using LAN bridging and switching techniques.

- WAN architectural design. This section will overview the different WAN topologies that may be used, and review common WAN technologies from which the designer might choose.

- WAN physical design. This section looks at the principal considerations the designer will have in converting the architectural design to a physical design that can be costed.

- Resilient WAN design. This section looks at strategies that may be adopted to increase the reliability of a WAN design, e.g. through use of a back up network.

- Legacy support. This section specifically addresses how we might accommodate non LAN attached devices into the enterprise network design.

- Gateway device selection and design. This section will look to selecting the appropriate gateway technology to interconnect LAN and WAN, and then look at what is required to design and dimension an appropriate device.

• Remote user and inter-organisation communications. Companies do not just need to communicate internally; they also need to be able to communicate with their customers and suppliers. This section looks at some of the possible approaches of adding this external communication requirement into the enterprise network design.

5.2   LAN DESIGN FOR SMALL SITES

With modern enterprise networks, a significant amount of design effort will focus on the local area network (LAN). LANs can best be understood by starting with the basic underlying topologies. All topologies start with the basic concept that all end systems connected to the LAN are sharing a common communications media. This not only allows simple any to any communications, but also permits an end system to broadcast to all other end systems on the LAN—a fact that is widely exploited in LAN protocols. These are shown in Figure 5.2.

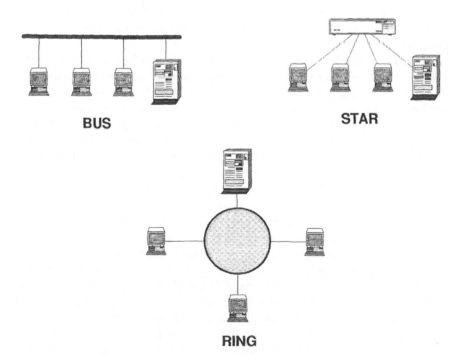

BUS

STAR

RING

Figure 5.2   The Basic LAN Topologies

- Bus. In this topology, the shared media is chained around the building, from end system to end system.

- Ring. This is wired in a similar way to the bus, but the media has to be deployed in the form of a closed loop.

- Star. In this topology, each end system is cabled to a central hub device. We will see that in practical LAN technologies, this hub is effectively implementing either a bus or a ring internally.

- Hybrid topologies. In practical LAN implementations, we can expect to see these topologies used together to create solutions. Examples will include buses linked by a star, sets of stars linked by a bus, stars of stars and rings to link rings.

Choosing technologies

For the majority of LAN based applications, the designer is likely to choose between one of four standards—IEEE 802.3 (Ethernet [Strictly speaking, Ethernet is a proprietary standard predating IEE 802.3. It uses slightly different voltage levels on the media, and different frame structures.]), IEEE 802.5 (Token Ring), FDDI or ATM. Each of these technologies has a range of variants, for both speed and media used. In all likelihood, practical designs (especially on large sites and campuses) will consist of a range of several technologies.

Ethernet generally provides the lowest cost interface adapters for the end systems (also, many models of end systems include Ethernet interfaces as standard). It can provide up to 10 Mbps of bandwidth. (However, due to the collision mechanism used for media access, this full bandwidth is seldom achieved. The exact amount of bandwidth available will be a complex function of the number of end systems on the LAN segment, their traffic rates and the size of messages they send. A safe assumption would be to plan for a LAN utilisation not exceeding 30%.)

Users requiring higher performance (e.g. those with bandwidth intensive multimedia applications) could consider using the 100 Mbps Ethernet standard (100BASET). Very high speed servers or host systems might be connected using one of the new Gigabit Ethernet standards (unlike other Ethernet standards that run over copper cabling, this is based on fibre optic).

Token Ring is widely used, though by no means exclusively, in IBM environments. New installations would be expected to operate at 16 Mbps (older Token Rings use 4 Mbps). The token passing media access method of this LAN type ensures that utilisation can approach 100%, giving a higher performance environment that Ethernet. Token Rings can also have better management capabilities than Ethernet (as there is always token traffic circulating, this can be monitored, and failure can be rapidly detected). The Token Ring technology also lends itself to more resilient large-scale LAN designs than can be achieved with Ethernet. The downside for Token Ring is the higher cost of associated equipment.

FDDI is an important niche technology. It is based on a dual fibre optic ring structure. It tends to be used on large sites to provide a backbone for interconnecting LAN on different floors or buildings across the campus. This is because it provides high speed (100 Mbps), large distances between stations and high reliability (due to its self-healing dual-ring structure). It is a comparatively expensive technology, and is not normally used directly by end systems, unless they specifically require the high bandwidth and high resilience this technology can provide. This technology is beginning to decline now, being replaced by fibre optic Ethernet and ATM technologies. It will not be discussed further in this book.

ATM started life as a WAN technology, but has been adopted for use in LANs. It is possible to design ATM to the desktop (usually using 25 Mbps ATM interfaces)—this is unusual however, as it would usually be more cost-effective to use 100BASET Ethernet. ATM generally finds its place in the backbone of large campus LANs, and its use here will be described in section 5.3.

The bus approach—10BASE5 and 10BASE2 Ethernet

Figure 5.3 shows a typical small site LAN as would have been implemented in the early days of Ethernet, using the 10BASE5 technology. The LAN centres around a thick coaxial cable running around the building forming a 'bus wired' topology. The cable has terminating resistors at each end (without these, signals on the cables would be reflected back along the cable at the open circuit end, degrading the quality of the signal). This span of cable forms what is known as a LAN segment. While this form of Ethernet is most unlikely to be deployed new today, it is important to understand it, as it will often be found as 'Legacy LAN', and much of today's LAN technology can be seen to be derived from it.

End systems will have Ethernet network interface cards (NIC) in them, terminating in a 15 pin 'AUI' interface. This is connected to an Ethernet transceiver via an AUI cable. With the 10BASE5 technology, the transceivers are clamped to the coaxial cable—they have a probe, which is made to penetrate the coaxial cable to make a contact with the central conductor and the outer screen. This arrangement is frequently referred to as a 'vampire tap'.

10BASE2 technology is a more recent variant on this theme. It uses less expensive thin coaxial cable, and exploits BNC connectors instead of vampire taps. To add a new station, it is necessary to cut the coaxial bus and terminate the two ends with BNC connectors. The new station is then inserted into the LAN using a BNC 'T' connector. The end system NIC

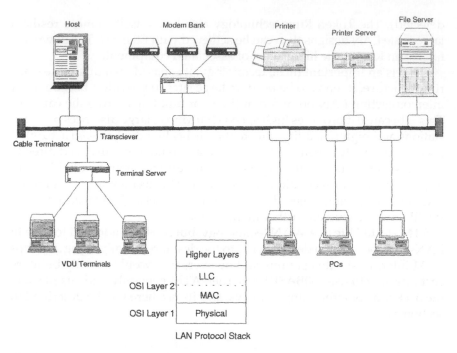

Figure 5.3 A Typical Small Site LAN

may terminate in an AUI—in this case a separate transceiver is required which provides the BNC connection to the Ethernet LAN. With the introduction of 10BASE2, however, it has become increasingly common for the transceiver to be integrated into the NIC, so that the Ethernet LAN is connected directly to the BNC connector on the back of the end system.

The main principle of Ethernet then, as now, is the concept of CSMA/CD—Carrier Sense Multiple Access/Collision Detect. If a station needs to send a message, it first listens to see if there is any traffic on the cable. If there is not, it is free to send. As utilisation of the cable increases (more stations are added or stations send increasing volumes of traffic), it becomes increasingly likely that two stations will elect to transmit simultaneously. Transmitting stations detect this by monitoring their own transmissions, which will be corrupted if a second station transmits.

If collision is detected, each transmitting station has to back off and wait for a short randomly selected period before trying again. Figure 5.3 shows the protocol stack used for Ethernet. The cable and the mechanism for sending and receiving a bit stream on it form the physical layer. Traffic is sent across the network as blocks or frames containing up to 1500 bytes of user data.

The CSMA/CD mechanism and the basic structure of these frames form the MAC (Media Access Control layer). A key feature of the MAC layer is that each station has a MAC address associated with the interface card (generally assigned by the manufacturer when the interface board is manufactured). A MAC frame will contain the source MAC address and the MAC address of the intended destination (destination may also be to the broadcast address—all stations on the LAN segment receive the frame).

Above the MAC layer is the LLC—Logical Link Control—this adds a header and checksum and allows for organised transmission of data between end systems. There are two forms of this. The most widely used is LLC1 which is a simple connectionless protocol. While this is able to detect and discard errored frames it relies on higher layer protocols such as TCP/IP or Novell IPX to provide reliable end-to-end communication. LLC2 is a connection-oriented protocol. It forms an association between two communicating end systems that provides reliable error corrected communication. It is most commonly found in an IBM SNA context.

In the earliest days of Ethernet, PCs were very rare—most workers would be using Visual Display Unit (VDU) terminals linked by serial links to mini computers or mainframes. Before the Ethernet, this would require large numbers of serial cables (which often had difficult length restriction problems to accommodate). With the introduction of the Ethernet, the host computer could have a single connection to the Ethernet. Located at strategic points around the building would be terminal servers. These interfaced the simple terminals to an appropriate LAN based protocol for carrying their traffic to and from the host machine. Example protocols used are the TCP/IP Telnet protocol and the DEC LAT protocol. A typical terminal server might connect 8 to 16 terminals to a single Ethernet transceiver.

As the PC became increasingly common, the Ethernet formed the ideal medium for interconnecting them. Each PC has its own interface card and is connected via an AUI cable to a transceiver. The PC used the LAN to communicate with shared file server or print server PCs. Another common use was to communicate with a modem bank (attached to a terminal server) for dialling out to early forms of e-mail service etc.

Extending the 10BASE5 LAN—introducing the star topology

The bus-based Ethernet LAN technologies have a couple of design issues which frequently cause difficulty for designers. The coaxial cable has a maximum run length (500 m for 10BASE2 and 185 m for 10BASE2) and taps can only be placed at a minimum spacing (2.5 m for 10BASE5 and 0.5 m for 10BASE2).

In order to extend the length of LANs, the LAN repeater was developed. This is shown in Figure 5.4. A Local Repeater is used to link two nearby LAN segments (e.g. segment A and B). The basic repeater is a two port device, with a connection into each of the two LANs to be connected. The repeater works at the OSI physical layer, receiving bits from one LAN

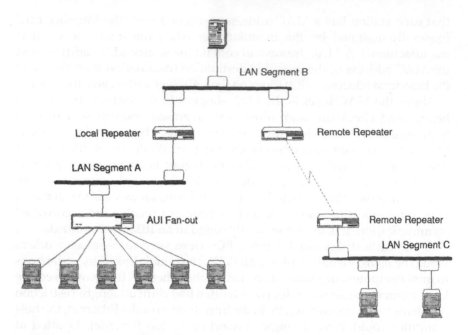

Figure 5.4 A Local Repeater Used to Link Two Nearby LAN Segments

segment and transmitting them into the other. LAN segments connected by a repeater effectively act as a single segment. Because of this, collision detection must operate properly across all segments linked by repeaters. The collision detect mechanism is timing critical, and as a result this limits designs using repeaters in such a way that signals may only ever transverse two inter repeater links (i.e. the signal can pass through a maximum of three repeaters).

If the LANs to be connected are not nearby, a pair of remote repeaters can be used. These can be used on large campus sites to interconnect buildings. The two devices are interlinked typically by a pair of fibre optic cables (or possibly by a line of sight laser links). The link between the two remote repeaters is counted as an inter repeater link. The remote repeater arrangement is shown connecting segments B and C.

In addition to the two-port repeater, it is also possible to obtain multiport repeaters that can be used to link multiple LAN segments into a star.

The 10BASE5 rule that devices must be connected to the media at distances of at least 2.5 m can be a real problem. This was overcome by using a fan out device. This typically would provide 8 AUI links for a cluster of PCs, and a single AUI link to connect to the LAN media transceiver. The fan out device can be seen providing service to a cluster of terminals on LAN segment A in Figure 5.4. It can be thought of as a nine-port repeater. The fan out effectively implements a star technology.

10BASET and 100BASET LANs

The early 10BASE5 and 10BASE2 LANs can be difficult to install (especially 10BASE5 with its thick media, which could only be bent around large radii without damaging the media!). They are also very inflexible once installed, making rearranging office layouts very difficult.

The use of star-based wiring is much easier to plan and much more flexible. A building can be flood wired with a cable from every point where a user is likely to be located, and taken back to a central point (often referred to as a wiring closet). This approach is identical to that used to wire phone equipment (indeed it is common practice to provide multiple cables to each desk position which can be used for either telephony or LAN connection.

10BASET and 100BASET LANs use this star approach. (Note that 10BASET had a predecessor standard simply known as UTP. UTP and 10BASET LAN segments use slightly different hub specifications. UTP and 10BASET hubs can be mixed on a common backbone LAN.) 10BASET and 100BASET LANs are generally implemented using high-quality UTP (Unshielded Twisted Pair) cables. This usually uses cables to the CAT5 standard, although higher grade cable can be used to 'future proof the design' (higher grade cables will be able to handle higher speed LAN standards that may emerge in the future, but will be considerably more expensive). PCs or other networking products using 10BASET or 100BASET will frequently have an integral transceiver, terminated in an RJ-45 socket. Building wiring is also terminated at an RJ45 socket. To connect to the LAN, a user will need an appropriate drop cable terminated with RJ45 plugs, in order to connect the PC to the building wiring connector.

Back at the wiring closet, all the separate circuits are usually terminated on a patch panel. This allows for maximum flexibility in that circuits can be assigned to a port on a specific LAN hub, patched through to a PABX or left unjumpered for future use. LANs are created by connecting circuits onto ports of a LAN hub. These behave very much like the LAN fan outs described for 10BASE5 LANs. The LAN hub will either be designed to support 10BASET or 100BASET. These technologies cannot be mixed in a hub (if mixing is required, LAN switches can be used—see section 5.3).

LAN hubs can be simple devices serving only to implement the LAN. Alternatively, they can be sophisticated manageable devices that allow network managers to manage access to the LAN and to monitor in detail traffic on the LAN (using RMON technology—see Chapter 7 for more detail). Whatever type of LAN hub is chosen, it is important to keep in mind that these devices inevitably need to be supplied with mains power. Typical 10BASET or 100BASET topologies are shown in Figure 5.5.

The first thing to note is that LAN hubs come in finite sizes. They will have a fixed number of RJ45 ports, one being required for each end system to be connected. There are a wide range of products with numbers of ports

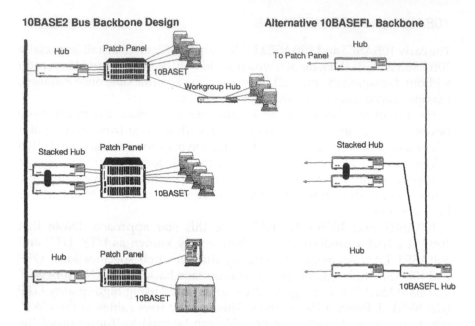

Figure 5.5 Typical 10BASET or 100BASET Topologies

typically varying from 4 to 24. Some products are described as 'stackable hubs'. This means that the number of ports can be expanded by stacking hubs and linking them with special cables that link together the internal buses of the units.

LAN hubs always have an 'uplink' port, allowing the hubs to be interconnected by a backbone LAN. This often uses a different technology than 10BASET, to allow greater distance between hubs. Typically 10BASE2 is offered (used to connect LANs as a bus). 10BASEFL is also sometimes used. This is a fibre optic implementation of Ethernet and offers good cable range, and freedom from concerns about electromagnetic emissions from the cabling or receipt of interference if the cable has to pass through electrically noisy environments. 10BASEFL units are star wired to a 10BASEFL hub. In addition to these approaches, it is always possible to use one of the 10BASET ports to provide an uplink to form a star of 10BASET hubs. A special cross-over cable will be needed to interchange the transmit and receive circuits (although many hubs feature a button associated with one of the ports that can be pressed to effect this cross over instead).

LAN hubs act as repeaters, and when designing with them, the 2 inter-repeater link rule should normally be observed. Rules will vary between different equipment however, so the designer will need to study the supplier's specifications. Typically, the links between stackable hub modules do not count as inter-repeater links.

Figure 5.5 shows two approaches to providing a LAN in a building. The ground floor has many users and needs a stackable hub. The top floor used a single hub but this has not proved adequate. It has been supplemented by a small 'workgroup' hub in a specific open plan area to support a cluster of users. The host and server systems are located in the basement. On the left hand side, we see that a 10BASE2 LAN has been installed in the building riser to interconnect the hubs. On the right, an alternative backbone approach using 10BASEFL has been used and an appropriate hub has been provided in the basement. Topologies in which a bus backbone is replaced by a star arrangement are sometimes referred to as 'collapsed backbone' designs.

Design summary for small Ethernet LANs

When designing for these smaller Ethernet designs, the designer will need to look at the following points:

- Which site types need small LANs? How many of each site type are there?

- For each site type, how many end systems are to be connected?

- For each site, how are the end systems distributed around the building?

- What will be the principal LAN technology? (Today, most designers would select 10BASET or 100BASET.)

- Will a single hub be sufficient? If not, how will hubs be connected together? (A star-wired design is likely to give the greatest flexibility for future changes.)

- What cable type will be used? UTP Cat5 is most commonly used, but many designers today will be considering higher grade cables to allow for future higher speed working. Shielded cable (STP) may also be considered (again, it will be designed for higher speed working, but offers added electromagnetic shielding, which may be important to address security concerns or to prevent possible interference with other equipment in a building). Clearly, the higher grade cables are significantly more expensive to buy—but the capital cost of cable is often small compared to the labour costs involved in wiring buildings, and the cost of staff disruption if the cable has to be changed in future!)

- What are the safety requirements? LAN cable is often affected by building fire regulations. Inexpensive cables often emit large quantities of toxic fumes in a fire and are thus not allowed. The designer may be required to specify more expensive cables which do not have this problem (often referred to as plenum grade).

- Should passive or manageable hubs be used? Often a mix of types will be appropriate, with passive devices for workgroups and active managed devices in the backbone.

- Which manufacturer should be used? There are many to choose from. Selection may be predetermined by company 'preferred supplier' arrangements. If this is not the case, we need to look at each manufacturer's range to see which will best fit our hub requirements, while providing the best overall price per port for the design.

In addition to the above, the designer must keep in mind the technical design rules associated with the LAN technologies being used. Key rules are:

- Cable distance. Maximum length of 10BASE5 cable is 500 m. The maximum length of a 10BASE2 cable is 185 m. The maximum distance from a 10BASET device to its hub is 100 m.

- AUI cables. Where an AUI cable is used between an end system and a LAN transceiver, this must be at most 50 m.

- Distance between transceivers. For 10BASE5 this is 2.5 m and 0.5 m for 10BASE2.

- Maximum transceivers on a single cable segment. For 10BASE5 this is 100. This drops to 30 on 10BASE2.

- Use of repeaters. A maximum of 2 inter-repeater links is permitted in any design. With 10BASE5, the maximum cable length between stations is 2.5 km when repeaters are used. With 10BASE2, this length is 925 m.

- Fibre optic repeaters. Maximum fibre length for FOIRL is 1 km and for 10BASEFL it is 2 km.

- Maximum number of end systems per LAN. When designing LANs, it must be remembered that a LAN is a shared medium. When one station is transmitting, no one else can. Therefore the absolute maximum throughput on a 10 Mps LAN is 10 Mbps! In reality, LANs normally have many stations wanting to transmit simultaneously. If the LAN is run to maximum utilisation, variable delays and packet loss caused by collisions will occur. It is wise to spend time considering likely traffic and ensuring that an Ethernet LAN segment is typically no more than 30% utilised at peak.

Token Ring LANs

The Token Ring LAN is implemented as a ring, see Figure 5.6.

Token Ring Logical Design   **Token Ring Implementation**

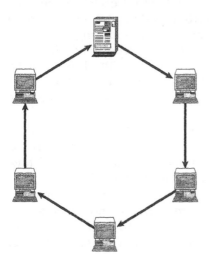
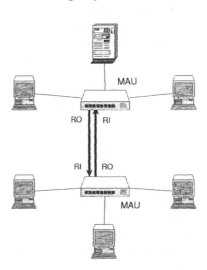

Figure 5.6   Token Ring LAN

User data is sent in frames, similar to those used with Ethernet. Unlike, Ethernet, the CSMA/CD mechanism is not used to access the media. Instead, a token passing mechanism is used. One member of the ring is designated as a master station. The master is responsible for injecting a token onto the LAN. This token is passed from machine to machine, until it arrives at a user who has traffic to send. This user replaces the token with a data frame, addressed to the MAC address of the recipient. The data packet then circulates around the ring until it is received by the recipient. The recipient will now replace the circulating data frame with a token, allowing another user to use the LAN. (This mechanism is known as 'early release'. An alternative method of working found in some Token Rings is for the packet to continue to circulate until it reaches the sender, who has the responsibility of removing it and replacing it with the token. This method is always used for frames sent to the broadcast MAC address, which have to be received by all stations on the LAN.)

Token Rings operate at either 4 Mbps or 16 Mbps. The frame size supported is larger than that for Ethernet (4 kbytes at 4 Mbps and 8 kbytes at 16 Mbps). Because this LAN technology does not suffer from the possibility of frame loss by collision, it can be designed to safely operate at much higher utilisations than an Ethernet segment.

In practice, Token Rings are implemented as though they were a star network. This is done by cabling end systems back to hubs, located in central wiring closets. These cable runs are referred to as lobes. In Token Ring terminology, the hubs are referred to as Media Access Units (MAU).

MAUs are designed so that they effectively implement the ring within the unit. To expand the Token Ring beyond a single MAU, each MAU features two special ports, referred to as Ring In (RI) and Ring Out (RO). RO ports are cabled to RI ports to link all MAUs into a continuous ring.

There are two types of Token Ring equipment. Type 1 is designed for installations using Shielded Twisted Pair cable (STP). Type 3 is for installations CAT 3, 4 or 5 Unshielded Twisted Pair cable (UTP). Both types can operate at either 4 or 16 Mbps. Type 1 cabling is more expensive, but longer cable runs are supported.

Hubs can be passive (non powered) devices, e.g. the IBM 8228 MAU or powered. Powered devices will generally support longer lobe lengths. There will also be a choice on managed and non managed LAN hubs depending on whether or not remote management is required.

Design rules regarding maximum length of cable runs and maximum number of end systems on a Token Ring are complex, and dependent on the design of MAU used. The designer will have to study the relevant manufacturer's technical literature. Typical 'rules of thumb' for Token Ring are:

- Maximum stations per LAN—260 for Type 1, 72 for type 3.

- Typical lobe length—100 m, but this is very dependent on type of MAU, cable type and ring speed. Repeaters are available, which are used to extend the distances allowed for cabling between RI and RO ports. These can allow up to 1 km on type 1 cable and 2.4 km on fibre optic. Repeaters may also be used to extend cable runs to single-end systems, where it is impractical to provide a local MAU.

5.3   LARGE-SCALE LAN DESIGNS

The key limitation with the LAN designs described in the previous section is one of bandwidth. The designs described operate as a single, shared, physical media and only one station on the LAN is able to transmit at once. The more stations on the LAN, the smaller share of the available bandwidth is accessible to each. Another key limitation of these LANs is one of distance. With Ethernet, the distance must not be too long, otherwise the collision detection mechanism does not operate correctly. With Token Ring, the delay of frames propagating around the ring would become excessive. Large-scale LAN designs must use equipment to overcome these limitations. To remove the physical layer limitations, this equipment operates at either the link or network layer.

Ethernet bridges

A typical bridged network is shown in Figure 5.7. This shows two types of bridge: the local bridge (used to link two LANs on the same site) and the remote bridge (used to link two sites).

Bridges operate at the link layer. These devices are generally easy to use, as they are 'self learning'. Imagine bridge A has just been installed. It will receive packets from LAN 1 and pass them on to LAN 2 and vice versa. As it passes packets, it observes the SOURCE MAC address and records this against the port on which the packet entered the bridge. In future, when the LAN sees a packet with that DESTINATION MAC address, it will know which port it is on. This means that if a packet has its source on LAN 1 and is destined to a MAC address found on LAN 1, the bridge *does not* need to forward it to LAN 2. As the bridge learns where each MAC address on the network is, it forwards less and less packets. This has two clear advantages:

• Unlike a design using repeaters, traffic local to one LAN segment is kept local. This reduces traffic on other LAN segments.

• In the case of remote repeaters, the bandwidth available to link them is often much less than LAN speeds. By only sending necessary traffic over the link, best use is made of the available bandwidth.

As well as working in self-learning mode, bridges may also be able to be configured with static MAC information. This can be a useful security

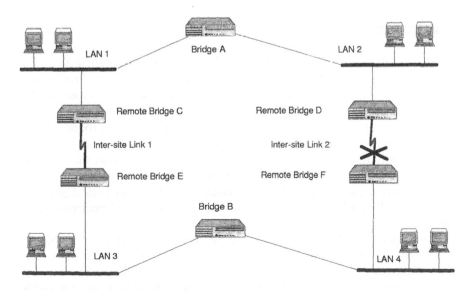

Figure 5.7   A Typical Bridged Network

feature, as you can limit which users are allowed to communicate via bridges.

In bridge designs, it is important to note that broadcast packets are always sent across bridges. This means that in the design in Figure 5.7, these broadcast packets could loop indefinitely around the bridges. To prevent this, the bridges run a protocol called Spanning Tree. Bridges send out their own messages aimed at identifying other bridges. They communicate together to establish the topology of the network. They will then 'prune' the network by shutting down bridge links, so that there is only ever one path between two points (i.e. no loops). In Figure 5.7, the large black X represents the link that Spanning Tree has currently shut down. Spanning Tree requires bridges to communicate every 2 seconds and this does use a fair amount of bandwidth. On the other hand, a faulty network connection is quickly discovered and the network can reconfigure by bringing previously shut down links into play. This means the design in Figure 5.7 has good resilience against failure of one of the inter-links or one of the bridges.

One useful feature of bridges is that they are able to interface between different types of LAN—for instance, LAN 1 could be 10BASET and LAN 2 could be 100BASET. This is because the bridge has to fully receive the frame, examine its contents and then resend it (unlike repeaters which work on a bit by bit basis). Bridges able to handle multiple speeds are generally able to automatically sense the LAN speed and self configure.

Switches

LAN switches are effectively multiport bridges. They are configured with (or learn) the list of MAC addresses attached to each port. Traffic only flows between ports when a destination MAC address is not local to that port. In a switched environment, LAN traffic is only present on LAN segments containing the source and destination addresses, freeing bandwidth on other LANs to be used simultaneously by other users.

LAN switches come in a range of sizes. Typical low end units would support 8 ports, with high end systems handling perhaps 100 ports. They can be used anywhere that a LAN hub can be used in a network design—in fact upgrading a hub to a switch is a common tactic for improving LAN performance.

The price of LAN switches tends to be higher per port that a LAN hub, so it may not be cost-effective to use a switch in all locations. In this case, a switch is generally deployed as a backbone device, with ports connected to LAN hubs, which in turn serve groups of end users. Where possible, one ensures that groups of users who communicate between themselves are placed on the same LAN segment, along with servers specific to that work group. Servers shared by multiple workgroups will generally be given their own port on a LAN switch. In these cases, where Ethernet is effectively point-to-point, a special full duplex form of Ethernet can be used. In this, the server can send and receive simultaneously, effectively doubling the available bandwidth.

LAN switches can be linked together into networks. Like bridges, Spanning Tree is used to prevent looping. Switches will usually have one or more ports designated as 'up links', used to connect to other bridges. These will use a high-speed LAN technology (100BASET or Gigabit Ethernet). Large campus sites may have ATM networks, and some switches provide ATM encapsulation on the up links, so that bandwidth from the campus ATM network can be used. (In this case, a technology called LAN emulation—LANE—is required, to establish the location of destination MACs. Switches connected to the ATM cloud have to register known MAC addresses with a LAN Emulation Server. They are able to look up the location of unknown MAC addresses on this, and forward the frame over an appropriate ATM PVC. Another server is needed to handle frame replication for broadcast packets.)

Virtual LANs

There will always be a limit to how many end systems should exist within one bridged environment. It must be remembered that switches, like bridges, do not control broadcast traffic. This broadcast traffic limits the size of switched LAN designs. In addition to this, switched LANs effectively offer an any to any connectivity. This can be undesirable for security reasons.

Many LAN switching products offer a solution to this in the form of Virtual LANs (VLANs). Each switch port is assigned to a Virtual LAN, and the switch treats each VLAN separately, with no capability for traffic, including broadcasts, to pass between them.

In order for VLAN technology to work, additional information identifying VLAN membership has to be added to frames passing over the switch up links. Special protocols are used for this—perhaps the most common is ISL (inter-switch link).

VLANs are important in flexible working environments where people frequently change office location (e.g. hot desking) or workgroup (e.g. working in virtual project teams). Changing switch ports between VLANs can be easily achieved by the LAN management system, and does not need rewiring at the switch.

Routers

When VLANs are used, there is generally a need for some traffic to flow between VLANs. There is frequently a shared server, for instance. Also, employees will need access to the corporate intranet for e-mail and Web servers.

Any connectivity between VLAN has to be achieved using the network layer protocol. If end systems are using IP, then each VLAN will need to be in a separate IP subnet (see Chapter 6 for an explanation of IP addressing schemes). A router is then used to route traffic between VLANs. Routers can be equipped with sophisticated filters to determine what traffic is allowed to flow between VLANs.

The router is connected onto a switch up link, and must be able to handle the ISL protocol to identify VLAN membership for each frame. Because the router has a single connection to the switch, it is often referred to as a 'router on a stick'!

Large LAN switches are modular in design, and can often accommodate a router card to carry out this function. (Note that the opposite can also apply—large routers are modular and can accommodate LAN switch cards!)

Token Ring bridges, switches and translational bridging

Token Rings also have their bridging technology. This uses a technique called source route bridging and is described elsewhere in the book. Similarly, this can be extended to provide a switched solution.

Many vendors are also able to provide technology to translate frames between LAN standards (e.g. between Token Ring and Ethernet). This is called translational bridging. The main concern here is the different frame sizes used by these technologies. Translational bridging only works if end systems can be set up to send frames that will fit with the smaller frame size (i.e. 1500 bytes for Ethernet).

5.4 WAN ARCHITECTURAL DESIGN

A key factor in deciding what kind of wide area technology to deploy will be the topology of the network. This can be understood as the general shape of the network, or which site is connected to which. (Note that, ultimately, we are interested in which sites need to communicate with one another. It is reasonable to expect those sites exchanging the highest volume of data will be directly connected to one another. It may not be cost-effective to interconnect all sites that need to communicate. Those exchanging only small quantities of data may be best served by interconnecting via intermediate transit sites.) We will now survey the most important connection topologies which the designer will encounter. The topologies are summarised in Figure 5.8.

1 to 1—point-to-point

This is the simplest case, with one site communicating with another. In the enterprise network, there are often two key host sites—a main and a standby. These would be interconnected by a high speed point-to-point circuit to keep the data synchronised on the two host systems.

Many to 1—star or hubbed network

This is an extended form of point-to-point, with many individual point-to-point circuits converging on to a central hub site. An example would be a

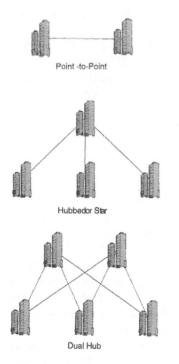

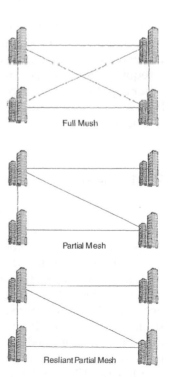

Figure 5.8 The Main Network Topologies

bank host site with connections out to its branches. The majority of application traffic in this case would be flowing between the branch and the central host. In this topology, if branches need to communicate with each other, they can only do so via the hub site.

Many to 2—dual hubbed network

This is a special extension of the many to 1 topology, which is worth mentioning, simply because it is so common. This is typically used where there are two host sites. The two sites provide resilience, as the application can be satisfactorily run using one site only (under normal circumstances, both sites could share the traffic, or operate in worker standby mode). Branch sites will therefore need an attachment to both sites to achieve resilience against host failure. Such a design also provides resilience against failure.

Many to many—meshed network

In this topology, every site is able to communicate with every other site. An example might be a mail order company with a dozen regional depots. Orders can be received and handled at any depot. If a depot does not have a required item in stock, the company needs to be able to contact every other depot to try to satisfy the order. Similar levels of traffic are expected between any pair of depots, so there is no obvious hub location for traffic to transit by.

Partial mesh

This is an intermediate situation between a star network and a full mesh. Imagine a company dealing in stock markets around the world. There are seven offices in different countries. The company headquarters is in London and all branches trade frequently on the London Stock Exchange. It is therefore natural for all branches to have connections to London. Dealers in Frankfurt regularly trade with Paris, so these sites are directly interconnected. Tokyo, Singapore and Hong Kong also have significant regional trading, and are thus interconnected. Trading between any offices that are not directly connected is carried out via the London office. If a clear pattern of trading were to develop between another pair of offices, say Frankfurt and Tokyo, the company could consider if it would be appropriate to directly interconnect them.

Resilient partial mesh

This is a partial mesh, but operating the rule that each site must be connected to at least two others. This allows a site to continue trading if one of its links is lost—traffic to the site connected by the failed link would have to pass via one or more transit sites. In our imaginary network, New York would not be able to trade if it lost its link to London. The network could be made fully meshed by providing New York with a second connection, for example to Tokyo.

Ring

This is simplest case of a resilient partial mesh, with a site being connected only to its immediate two neighbours, with all sites being linked into a ring. This is the topology used by Token Ring LANs. In the wide area, it is an important topology for campus and metropolitan area networks, where sites can be linked using technologies such as FDDI. The modern high speed point-to-point services based on SDH (or SONET) are also implemented as rings, to provide increased resilience.

Hierarchical networks—mixing topologies

Figure 5.9 shows a hierarchical network. This is typically required in very large networks. The network is divided into regions. Customer sites in that region are connected to a concentration site via a regional network. The concentrators and host sites are interconnected via a backbone network. Some network designs may call for more than two layers (large public data networks are commonly designed using three or four layer hierarchies).

The technologies used in the regional network and backbone network can be different, provided the concentrator can handle the transition between the media types. The differences in WAN technology at the different layers will reflect the data rates required at the different levels. The large-scale example, Data Bank, discussed in Chapter 10, will use a two-level hierarchy of this kind.

Selecting wide area technology

At this stage, some fairly broad design work will often be done to select candidate technologies for the solution. 'Thumb nail' designs will be

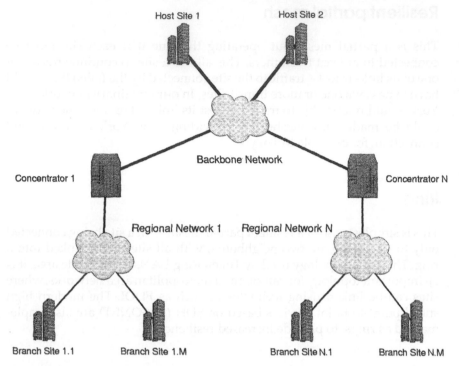

Figure 5.9 A Hierarchical Network

done using a range of technologies (e.g. digital private circuits, Frame Relay, SMDS). The design work will establish which technology is likely to have the most appropriate performance, availability and reliability. The designer will also want to compare appropriate solutions on a cost basis, by getting potential suppliers to provide budgetary quotes for some sample connections. As described above, in the section on hierarchical topologies, it is *not* necessary to adopt a single technology solution, and the best designs will often combine a range of technologies.

Before choosing technologies, it will be necessary to find out exactly what technologies potential suppliers have on offer, what their technical specifications are, and (often very important) what the tariffs are.

WAN technologies come in a variety of 'flavours' depending on the level of protocol support provided by the network. In terms of the seven layer OSI network model, protocols are offered that support the physical layer (simple point-to-point bit stream services), link layer (e.g. Frame Relay, offering basic frame delivery and multiplexing services), the network layer (e.g. IP VPNs, providing value-added services such as switched connections, multiplexing and quality of service) and application layer services (providing specific support for applications such as e-mail and EDI).

When deciding what would make a suitable technology, the designer will ask a number of questions and see how the answers match up for a range of available technologies. There follows a typical list of questions and some answers based on the following available technologies:

Physical layer services

- PSTN Dial with modems
- ISDN dial with terminal adapters
- GSM modems—for cellphone based mobile communications
- X.21 switched services (or switched 56 k service in the USA)
- Digital point-to-point.

Note that, while PSTN and ISDN are primarily used to set up physical layer bit paths, they can also be considered network layer services, due to the ability to set up switched calls to any number of locations. ISDN can also provide simple multiplexing by the use of multiple B channels. ISDN D channel can also be used to provide a packet based multiplexed service, based on X.25.

These services provide a basic end-to-end bit transport service. The users end systems will need to apply higher level protocols to encapsulate the application data and ensure reliable transport across the network. Typically, the HDLC protocol is used to frame data and provide basic error detection. TCP/IP is widely used to provide reliable end-to-end transport for the application data.

Link layer services

- Frame Relay

This type of service takes user data and transports it as frames (based on the HDLC protocol). It can offer basic multiplexing, with the link layer address being used to identify a specific destination to be reached over a preconfigured network path (known as a Permanent Virtual Circuit—PVC). This type of network can be referred to as a Non Broadcast Multiple Access (NBMA) network—frames sent on a connection can be sent to multiple destinations, but the network does not have the ability to broadcast. If an end system needs to broadcast to all destinations, it has to achieve this by copying the packet and sending individual copies to each destination. Frame Relay services also provide basic quality of service and congestion control.

Note that a Frame Relay service always has to be delivered over one of the physical layer networking technologies mentioned above.

Network layer services

- X.25 Packet Switched Networks

- X.28 Packet Assembler Disassembler (PAD) services associated with X.25 networks

- SMDS

- ATM

- IP VPNs.

In network layer services, each end system will have an associated network layer address. Packets are delivered to the specified destination address, offering the potential for highly flexible 'any to any' communication. These services may either be connection oriented (where a session has to be established with a required destination before packets can be exchanged, e.g. X.25 and ATM) or connectionless (where each packet is routed individually, e.g. SMDS and IP).

X.25 is not generally suitable for LAN interconnect (due to low bandwidth supported and relatively high network delay), but it is an important legacy protocol which networks may have to deal with.

SMDS and ATM are widely used for high-speed LAN interconnection. IP VPN is an emerging technology based on Internet technology and direct transport of IP packets. Unlike the public Internet, IP VPN services create a 'Virtual Private Network' for each customer, but over a shared IP switching platform.

Note that network layer services always have to be provided via a physical layer bearer service.

Application layer services

- Internet (when the network supplier is providing value added functionality such as e-mail post boxes or World Wide Web servers). It could also be considered a network layer service if simply delivering IP packets between customer end systems.

- X.400—e-mail post office facilities where international ITU recommendations are being followed.

- EDI—Electronic Data Interchange—the network will provide store and forward mail boxes for handling the EDI structured business forms, allowing them to be exchanged between the host databases of customers and their suppliers.

Note that these value-added application support services always have to be delivered via a physical or network layer service.

Table 5.1 A Guide to Suitability of Available Technology

Technology	Speeds	Notes
PSTN Dial	300 bps–56 kbps	Top speeds will depend on voice quality of phone services used. In some locations and for long distance international calls, speeds above 2.4 kbps may prove difficult. Higher speeds can be achieved through use of data compression.
ISDN Dial	64 kbps–2 Mbps	A basic ISDN call provides a 64 kbps synchronous connection on a B channel. For higher bandwidths, multiple B channels can be 'bonded' together. A basic rate ISDN service provides 2 B channels and a primary rate service provides 30 (24 in the US). Note that if asynchronous data is to be sent over the service, this is adapted using V.110 or V.120, and the maximum rate is then 38.4k bps.
X.21 or Switched 56 Digital	64 kbps or 56 kbps	
point-to-point	9.6 kbps–Gbps	Typical service for branch locations might be 64 kbps and larger host sites would use 2 Mbps (E1) or 1.5 Mbps in USA (T1). Higher speeds are generally supplied using fibre optic cables. Increasingly, high-speed services are delivered using SDH or SONET technology which can provide resilient networking as these services are implemented as self-healing rings.
X.25	2.4 kbps–2 Mbps	Not widely used for LAN interconnection due to limitations in throughput, packet size and network delay.
Frame Relay	64 kbps–2 Mbps	Networks increasingly supply gateways to ATM, allowing interconnection to high speed sites. The designer needs to select access circuit speed and throughput parameters for each PVC.
SMDS	192 kbps–25 Mbps	In some services, the higher speeds may only be provided as point-to-point connections. Any to any working is generally available to 10 Mbps.
ATM	2 Mbps–2.4 Gbps	The designer needs to select an access circuit speed and then decide throughput parameters for preconfigured PVCs. Bandwidth for switched calls has to be negotiated at call set-up time.
IP VPN	64 kbps–Gbps	This is still a relatively new service, with many networks still at the planning stage.

Questions to ask when selecting WAN technology

What bandwidth is required?

Clearly, the network technology to be used must be able to support the peak data rates demanded by the application. Available bandwidth varies widely with technology. Typical values are discussed in Table 5.1 (but expect them to vary between service providers!).

What size packets have to be sent?

Typically, when interconnecting LANs, the user will want to be able to send packets intact. The largest packet size that any network element will support is referred to as the Message Transport Unit (MTU). If a packet larger than the MTU has to be sent, then a fragmentation method such as IP fragmentation has to be employed to cut the packet into suitable size pieces. The extra processing required and extra bandwidth needed for fragment headers make this undesirable.

Ethernets have an MTU of 1500 bytes. One reason X.25 is seldom used for LAN interconnection is that it has low MTU size—typically between 128 bytes and 1024 bytes. Fragmentation is, therefore, frequently necessary. Frame Relay and other high-speed services have MTUs of 4 kbytes and above and are therefore suitable.

Special care is needed when working with high speed Token Rings which can have MTUs of 8 kbytes. If the network will not transport packets this large, it is often possible to tune end applications to use smaller packet sizes, so avoiding fragmentation in the network.

If the network is designed using physical layer services, the question f network MTU clearly is not an issue.

Is bandwidth requirement constant or bursty?

The physical layer services provide a constant bandwidth, available at all times, whether the user needs it or not. Because the bandwidth is always there (and being paid for!), this service may not be cost-effective if the data is bursty. This type of service can be replicated on ATM networks if a Constant Bit Rate (CBR) connection is established.

The link and packet layer services are based on frame (or cell) multiplexing. Users share core bandwidth. Network designers assume that user data is bursty and adopt a contention ratio in the design. Put another way, there may not be enough bandwidth in the core if everyone wanted to send at once but, based on expected statistical distribution of user bursts, there is enough for the great majority of the time. In these networks, there must always be some mechanism of ensuring equitable distribution of available bandwidth (Quality of Service) and a mechanism for protecting the network from overload.

In general, the Quality of Service aspect is carried out by having each network connection state in advance the minimum amount of bandwidth it needs (a Committed Information Rate—CIR). The network will endeavour to ensure that this bandwidth is available for the great majority of time. In addition to the CIR, users will be allowed some ability to burst above this level. Congestion control is done by simply discarding excessive packets in the network. Burst traffic above CIR is generally marked as such and discarded preferentially.

The designer has to be aware of the possibility of discard in the network. CIR has to be designed to minimise this (traded off with the fact that higher CIRs will cost more!). Also, a reliable end-to-end transport mechanism (e.g. TCP) must be used to recover from discards.

How frequently is the network to be used?

The majority of WAN technologies discussed here have permanent connection to the network. The service is available 24 hours a day, whether it is used or not. The cost of the service will reflect this. If the service is only required for short periods in the day, these services may not be cost-effective. Examples would be the home worker, logging on to the network a few hours each day to do e-mail or access to the corporate intranet. Also, shops may need an hour on-line each night to transfer information about the day's sales. During the day, it may need to make a number of short on-line transactions, e.g. to authorise credit card transactions.

In these cases, it is often worth considering use of ISDN (or perhaps PSTN) as the primary WAN solution. Costs for this service are generally time dependent. If we plot a graph of costs against connect time, there will be a breakeven point below which an ISDN solution will be less expensive than a dedicated WAN service. The designer will have to carefully estimate likely connect time and find out how it relates to the breakpoint for available ISDN and dedicated WAN services. If an ISDN solution is selected, careful monitoring will be required to ensure use remains below the breakpoint. If it unavoidably exceeds the breakpoint, it will be worth migrating affected sites to a dedicated solution.

Further design issues for WAN services

Point-to-point digital leased circuits

The price of these circuits generally carries a high distance-dependent element, and design work will aim to keep the cost of these circuits to a minimum. A partial mesh is the usual solution, with resilient partial meshing for the most business-critical sites (less critical sites can other be non resilient or backed up using ISDN.

Frame Relay

Where Frame Relay is to be used, an added complexity is that current services are built around Permanent Virtual Circuits (PVCs) set up between two points at subscription time. Each PVC is assigned a Committed Information Rate (CIR) which specifies the amount of bandwidth to be allocated to that PVC. (Bandwidth management in Frame Relay networks is a complex area, and the precise meaning and implementation of parameters varies between technology platforms and network providers. The designer should clarify with the network supplier exactly what is being offered.) For each site type, the designer has to decide what PVCs there will be and what bandwidth to allocate to each. This information is conveniently represented as a From/To matrix.

A network supplier will have rules governing the total amount of CIR allowed on a connection of given bandwidth. A naive approach is to assume that the sum of CIRs on a circuit should equal its bandwidth. This is very restrictive, and fails to make allowance for the bursty nature of the traffic flowing over the PVCs. Network suppliers will, therefore, permit the sum of CIRs to be a multiple of the access circuit bandwidth. This is known as over-subscription—for example, a factor of 2 may be enforced.

SMDS

SMDS is a strictly connectionless protocol, and there is no concept of assigning a given throughput between two locations. The design is only constrained by the rate that data can be sent into the network (determined by the throughput class) and the rate at which data can be received from the network (determined by the access circuit speed). A site can address and send a packet to any other site connected to the network—so an SMDS network can effectively be regarded as a fully meshed network. SMDS also supports security features to restrict who can communicate with whom. This is done by defining address screens at the point of entry into the network. These list the addresses with which a site may communicate. Typically, a company would use these screens to set up a closed user group, isolating its traffic from other users of the SMDS network.

SMDS also supports multicast groups, and with these, an SMDS network can behave similarly to a LAN. When a site sends a packet to a multicast address, the data is delivered to all members of the multicast group. The multicast group is fully meshed. It should be noted however, that many SMDS networks restrict the size of individual multicast groups.

ATM

This service is often used to provide high speed cores in hierarchical network topologies. Network suppliers are increasingly providing gateways between Frame Relay services and ATM networks, so that lower speed sites can attach using Frame Relay, yet have traffic delivered to the high speed core sites (e.g. the main host/server sites).

ATM today is principally provided using PVCs. As with Frame Relay, there are quality of service parameters to be designed for. ATM typically supports three classes of service. These are:

- CBR (Constant Bit Rate). This guarantees a fixed end-to-end bandwidth with low and relatively constant delay (low jitter). This is ideal for private circuit emulation, when carrying high quality voice or video traffic.

- VBR (Variable Bit Rate). This is similar to Frame Relay, with a committed bandwidth and burst parameters. ATM networks strongly police conformance to the agreed parameters and any excess traffic is discarded. It is therefore essential that routers using ATM have appropriate traffic shaping capabilities to ensure traffic sent to the WAN is within the agreed parameters.

- ABR (Available Bit Rate). This is similar to VBR, but there is no committed bandwidth. This can be used for low priority/non time-critical traffic, such as that between e-mail post offices.

IP VPN

These services use Internet technology, switching exclusively IP traffic. Unlike the Internet however, traffic from individual companies is kept separate within a Virtual Private Network. This is achieved using a core of Gigabit speed routers, operating a tag switching protocol called MLPS. On entry to the network, a tag is added to each packet to identify its VPN. A routing decision is made to identify the destination router, and a tag added to identify the network path to take. The packet is rapidly switched through the network based on the routing tag (rather than by processing the destination IP address). At the destination, the VPN is checked, and if the destination port is in the same VPN, the packet is delivered.

As an IP service, a router is usually supplied at the customer's premises, and the service interface is an Ethernet port into which the customer LAN can be connected.

IP VPNs will support Quality of Service based on the QoS field in the IP packet header. It will be up to the customer's end system to mark each application with an appropriate QoS field value (this can also be set or enforced by a router). There are expected to be four classes of service—the highest will give guaranteed delivery with low delay and will be used for voice over IP traffic. The lowest might be used for e-mail delivery.

The designer will need to identify how much traffic there is likely to be in each class of service and agree a throughput for each with the service provider. Any traffic above this rate within a class is marked as 'out of contract' and will be preferentially discarded by the network if there is a risk of congestion.

5.5 WAN PHYSICAL DESIGN

The key to designing the inter-site or wide area network is to reach a good understanding of how traffic flows to and from each generic site type. The most important factor is to establish the peak flows, and understand how this may grow with time. The bandwidth requirement from each site can be a key factor in choosing WAN technologies, as each technology tends to have a particular bandwidth niche. For example X.25 typically serves 2.4 kbps to 64 kbs, Frame Relay serves 64 kbps to 2 Mbps and SMDS serves 2 Mbps and upward.

Establishing the peak bandwidth requirement for a site may not be easy. For existing legacy systems that must be supported over the enterprise network, good statistics may be available or measurements can made. A much poorer understanding is likely concerning peak loads for new applications, for example e-mail. How many mails will an individual receive and send at a session? When will the peak usage time be, and how large will mails be? Remember that with modern e-mail packages it is all too easy to attach a multi-megabyte file with just a few clicks of the mouse! This is increasingly the case as multimedia applications take off. Pictures can be hundreds of kilobytes and audio or video sequences many megabytes in size.

There is a strong possibility that the requirements gathering process has not provided as much quantitative information on traffic loads as would be liked. This can be specially true when considering traffic growth requirements (a typical network supply contract will run for five years— accurately forecasting traffic over this period will be next to impossible).

The solution here is for customer and network supplier to agree an initial bandwidth provisioning based on best available information, and clearly document the agreement in the contract. Customer and supplier should ensure the contract allows cost-effective speed and technology upgrades during the contract life.

Where traffic levels have not been provided in sufficient detail, the designer must adopt a 'rule of thumb' approach to the design, estimating traffic volumes for each application, agreeing these assumptions with the customer. When these assumptions are made, a formal issue should be recorded requiring these assumptions to be verified during service trials. Some examples of transaction profiles are to be found in the latter part of this book, where a detailed design case study will be undertaken in Chapter 10.

In estimating traffic levels, we must not forget to make allowances for protocol overheads. For example, as shown in Figure 5.10, TCP/IP packets

Frame Start Flag	1 Octet
Frame Relay Header	2 Octets
Protocol Encapsulation	3 Octets
IP Header	20 Octets
TCP Header	20 Octets
Application Data	N Octets
Frame Check Sequence	2 Octets
Frame End Flag	1 Octet

HDLC 'bit stuffing' which hides any flag patterns in the frame contents adds further overhead - 1.6% on average.

Figure 5.10   Protocol Overheads

passing over a Frame Relay network will typically have 6 bytes of Frame Relay overhead, 3 bytes of RFC 1490 protocol encapsulation overhead, 20 bytes of IP header and 20 bytes of TCP header. In the worst case, each packet could only contain 1 octet of user data (e.g. Telnet traffic from a terminal).

Also, allowances must be made for overheads created by LAN protocols and routing protocols. Examples here would be Novell's SAP (Service Advertisement Protocol) and the TCP/IP RIP (Routing Information Protocol) Both these broadcast periodic updates across the network. Network management traffic can also be a significant overhead. It is not uncommon for these non application overheads to run to 5–10% of the bandwidth requirement!

When selecting the speed of connection for a given site, the upper and lower bound will be set as follows:

- Lower bound. This must obviously support the peak bandwidth required by the application traffic transmitted by and received at this site.

- Upper bound. Put simply, the customer should buy as much bandwidth as can be reasonably afforded. Obtaining bandwidth higher than the lower bound will provide protection against poor traffic level assumptions and capacity for future growth. The key advantage will be in terms of performance. The access circuit from a branch into the WAN is generally the slowest component of the enterprise network, and can contribute the majority of network delay. Using increased bandwidth directly reduces this delay.

5.6 RESILIENT WAN DESIGN

The most common point of failure in any enterprise network is generally the physical circuit connecting a site into the core data network. These circuits are prone to damage by the natural environment (e.g. water damage to cables) and also human activity (such as errant road diggers!). Service may also be affected by failure of the network terminating units, or of the core network node to which the site is connected. It is not untypical for such problems to cause one or two service outages per year, resulting in loss of perhaps eight hours usage. For a business-critical application, this is unlikely to be acceptable, and the design must provide increased resilience. Naturally, this will increase costs! A number of techniques exist to do this.

Access circuit dial around

In this technique, the core network supplier will provide a mechanism for using the public telephone network to divert around a failed access circuit. In the case of enterprise networks, the speeds involved usually call for ISDN, although the technique is also used with analogue modems. A typical set-up is shown in Figure 5.11. In this, the site router is connected to an ISDN Terminal Adapter. This normally passes traffic straight through to the digital access circuit network terminating unit.

However, should the network terminating unit signal failure of the access circuit, the ISDN Terminal Adapter will make an ISDN call to a peer Terminal Adapter at the core network port. This approach is often the cheapest form of resilience, but it has its limitations! Key issues with it are:

- It does not protect against failure of the Terminal Adapters or core network port.

- The ISDN and digital access circuit probably share the same cable as far as the first local exchange—so if this cable is cut, the resilience mechanism will not work.

- The ISDN calls can be expensive—special care must be taken to ensure that management systems are aware that a site has switched to ISDN

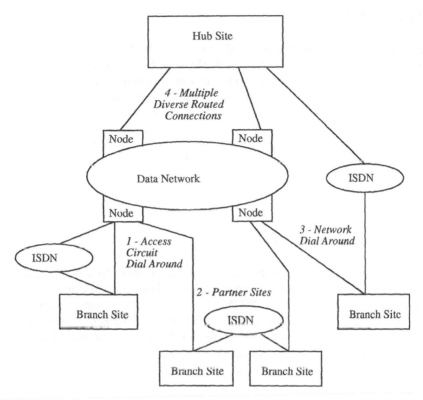

Figure 5.11 Typical Dial Around Set-up

working, and careful testing is needed to prove that the ISDN call is properly terminated when service via the digital access circuit is restored. A neat trick here is to negotiate for the network supplier to pay for the ISDN calls. Network connection charges will increase by a certain fixed amount assuming the costs of agreed connection outage times. The network operator will now be automatically 'punished' for any excessive periods of connection failure.

- Security can be a risk. Fraudulent use of the ISDN dial back up path can be prevented either by using Calling Line Identification at the core network terminal adapter, or by having the core network terminal adapter originating all back up calls via incoming calls barred circuits.

- In its simplest form, this approach can only provide a 64 kbps back up path. If higher speed service is to be protected, this may be possible using multiple ISDN B channels in conjunction with a channel aggregator (also known as an inverse multiplexer).

Partner sites

In networks with large numbers of branches, dialled back up may be performed by pairing off branches—ideally selecting branches that do not connect to the same network node. If one branch loses its network connection, it can then dial to its partner, and share the network connection of the partner. Both sites may of course, suffer some degradation of service if the back up path is initiated at times of peak usage. The approach can be attractive in that ISDN connections are only needed at the branch sites and not in the network or at host sites—thus reducing ISDN line rental costs.

Network dial around

Many large enterprises require branch sites to connect to a host/central server site. A useful approach for resilience in this case is to have branches with failed connections make ISDN calls to a hunt group of terminal adapters at the host site. This approach has the advantage of protecting against the failure of the network node to which the branch is attached. Such an approach is clearly not useful to an enterprise network requiring sites to have connections to many destination sites.

Multiple site connection

Perhaps the most satisfactory way to provide protection from failure of a circuit is to provide a second circuit—where technically possible, this should be provided via a diverse route (separate cable routes to different network nodes). These circuits may be used as a worker/standby system, or may share load under normal conditions. Multiple site connection is an expensive solution, and is normally only justifiable at central host/server sites, where loss of service would impact many other branch sites.

Where a business relies on services from a central host or server site, it is usual for there to be two separate host sites at different geographical locations. Either host should be able to deal with all traffic if the other site is unavailable. Under normal circumstances, these hosts can be either load sharing or worker standby. When designing the network, it will be necessary to ensure that branches are able to connect to either host site. Tie connections between the host sites are normally required to allow the hosts to keep there data synchronised.

5.7 LEGACY SUPPORT

Finding the best way to support legacy systems over an enterprise network based on LAN interconnect is not without its difficulties. Each legacy system to be supported must be examined in turn, and an appropriate method identified. Four major strategies exist to handle it:

- Scrap. It may be that a legacy application and the related end system hardware is nearing the end of its service life anyway. It may be possible to phase plans for replacement of the application with deployment of the new enterprise network. When re-implementing legacy applications, it is often natural to look toward distributed client–server applications, with client software running on LAN connected PC platforms. These systems can thus be designed to fully exploit the proposed enterprise network.

- Trap. We can decide simply to leave it alone for replacement later. It will be necessary to retain a certain amount of legacy hardware and network to support these applications. The main requirement here will to contain or trap the application, keeping its costs to a minimum. Development and deployment should be frozen, and the system withdrawn wherever and whenever practical.

- Wrap. Perhaps the most common technique is to retain the legacy systems, but find a way for the new enterprise network to transparently interconnect the legacy equipment (by wrapping their protocols in those of the new network). Examples of this technique are discussed in detail in later chapters.

- Map. Host systems may be upgraded to fit in with the new enterprise network architecture. However, there may be a large number of expensive end systems running legacy protocols which need to communicate with the host. Alternatively, the customer may have rolled out LAN attached PCs to users, and these need to attach to a legacy host system. In a mapped solution, we need to use a device able to translate between the legacy protocol and that used by the new system.

There may be occasions when it only makes commercial sense to install a new enterprise network if all legacy systems can be supported over it. What happens in the case that one of the legacy systems has a physical interface or protocol that cannot be supported by any known gateway device? In this case, it will be necessary to develop a custom device to handle the system. This may be done by the gateway manufacturer or by a third-party specialist company with the ability to write custom code for the gateway product. Such a development is usually only practical if the customer has sufficient investment in the legacy system to make the costs of the development attractive, compared to the option of replacing the legacy system. The timescales for developing and testing this custom gateway will have to be allowed for in the project plans. With any such development, there is a risk that it will fail to deliver on time, on cost or at all! Contingency plans must be in place to manage this risk.

Most types of legacy system will use a serial communications interface such as RS 232, V.35 or X.21. This can be cabled to the gateway equipment in a straightforward manner in most cases. Difficulties can arise if the legacy equipment is placed some distance from the gateway device. Serial interface technologies can have significant distance constraints, and it may be necessary to use line driver devices to extend the reach. Variations in interface standards (gender and type of connectors and pin usage) may also require the design and use of adapter cables.

5.8   GATEWAY DEVICE SELECTION AND DESIGN

Having decided on site LAN design, the legacy systems to be connected and the wide area connections to be supported, we are now in a position to select the type of gateway device. This is used at each site to interconnect the LAN and WAN. There are three principal types of device to consider:

- Bridges. These have been described in section 5.3. Bridging has to be used if the end systems use non routable protocols. The major problem with bridges is that they do not scale well. There are often severe hop count limitations and the Spanning Tree algorithm is quite bandwidth intensive. A big problem with large bridged networks is that of broadcast storms. A broadcast packet is always propagated to every end system on the network—so as the number of end stations grows, the amount of bandwidth taken up by broadcasts rapidly increases. Peaks in broadcast traffic can severely disrupt user applications.

- FRADs (Frame Relay Assembler Disassemblers). These are specialist devices used with Frame Relay networks. For LAN traffic, they act as bridges, encapsulating LAN frames into Frame Relay frames. They will also support HDLC based legacy protocols (e.g. IBM's SDLC)—again by encapsulating the frame into Frame Relay frames. Some FRADs have voice support capability, allowing analogue telephony connections to be digitised, and voice samples sent as frames.

- Routers. The majority of LAN interconnect solutions use these devices. By operating at the network protocol level (e.g. IP), they are able to eliminate most broadcast traffic (as this is largely associated with link layer functionality). Routers are thus protocol sensitive, and the designer needs to be sure that the router can support the appropriate LAN protocols. Where a protocol cannot be routed (e.g. NetBios, which has no network layer), routers often also support bridging functionality. The remainder of this section is devoted to helping the designer select an appropriate router for each site.

Routers are versatile but very complex devices. They are generally sold as basic units with many add-on options to be specified. To help ensure you get it right, suppliers will often provide 'worksheets' to help the customer fully specify the router. The following areas should be typically considered:

- What type and number of LAN and serial ports are required? This will help to decide if a low end 'branch router' (which has fixed port configurations) or a high end multicard router is needed (in which case you will have to select which LAN and serial interface cards are required).

- How much processor power is required? Most models of routers are highly optimised to switch IP datagrams. This is often carried out, or at least supported by dedicated hardware. If the router is asked to do anything other than IP switching (examples would be filtering traffic to provide a security 'firewall' or running a TCP/IP session to encapsulate legacy data), then that task is carried out by the routers processor. If asked to do large amounts of processor intensive work, the router can run out of processing power, and fail to deliver the data throughput required of it. If this is likely to be an issue, then you will need to look at routers with more powerful processors (an example is the Cisco 4500 which features a 64 bit 100 Mhz RISC CPU—this has twice the processing power of the Cisco 4000, but is, of course, more expensive).

- What software level and version is required? Software for routers is normally priced separately from the hardware. There will be a range of different levels to choose from. A basic level will typically support IP routing only. A medium level might include IP, Novell support, DEC Net and bridging. The top level will offer a comprehensive range of support for routable protocols, protocol translation and legacy protocol support.

Router software is also supplied in different versions. A router manufacturer will typically release between two and four major versions of software each year—each new version supporting a wealth of new features. Between major version releases, there will be a stream of maintenance release versions, fixing bugs found in the software. Some tips for selecting a software version are:

  - If adding routers to an existing network, get the same version as the other routers as forward and backward compatibility cannot be guaranteed.

- Use a recent version, as manufacturers rapidly cease to support earlier versions of code.
- Do not be in too much of a hurry to take the latest major version release, even if you are keen to exploit some of its new features. Router software is highly complex, and experience shows that early releases of a major version can have bugs—not only in new features, but often in basic router functionality. Thorough lab testing of such software is recommended before deployment, to confirm that the code is stable, and any new features to be used behave as expected.
- A good compromise for the conservative network designer is to select the latest maintenance release of the previous major code version.

- Memory. A router typically has two types of optional extra memory: 1. RAM, used during router operation for traffic buffers and other router data structures, and 2. non volatile memory, used for software storage. The general advice would be fit as much of both types as possible. This option is expensive, but as router software evolves, and the user becomes more demanding on the router (in terms of data throughput and the range of functions required) then more memory is required. The cost of revisiting several hundred sites to retro fit memory would be hard to bear financially, and there will also be outage of service.

Suppose the router only has enough non volatile memory to hold the operating software. It is likely that at some stage a new version of code must be downloaded to the router—to fix a bug or add new features—or simply to ensure that the code being run is still supported by the supplier. In this case, the router must stop normal operation, and will run code from a boot ROM. This will be of sufficient complexity to allow the router to connect to a file server in the network and download the new code image using a file transfer protocol (typically the TCP/IP TFTP protocol). The downloaded code overwrites the previous version. The router will be out of service for the time it takes to download the image (perhaps 4 Mbytes) and restart the router running the new code. To contain such outage, Non volatile memory should be at least twice as large as the software image. This will allow a software upgrade to be downloaded to a router which is still in service, minimising down time.

5.9   INTER-ORGANISATION COMMUNICATIONS

Inter-organisation connectivity

All our discussion above has focused on the need for applications to communicate within the organisation. Of course, most companies will have some need to communicate with other organisations. This might be with a specific trading partner or just the non specific ability to communicate with the wider world (as might be typified by connection to the Internet). A further key form of 'inter-organisation' communication is where a company may wish to offer its services electronically to the general public (via Internet or other dial access networks). When considering this form of connectivity, key points that have to be considered are:

* Cost containment—unrestricted access to inter-organisational networks can prove expensive.

* Security—there is invariably a need to control traffic out of and more particularly, into the corporate network.

* Compatibility—are the corporate applications in fact compatible with systems and networks owned by other parties with which the company needs to communicate?

Selection of an appropriate interconnection method will help to address these issues.

Interconnection technologies

Obviously, to be able to intercommunicate in the first place, the organisations must be attached to networks that will allow the intercommunication to take place. Some technologies are better at this than others!

Private circuit

Two organisations wishing to intercommunicate could arrange to link their sites by private circuit. This is commonly done where the need for control and security is high (e.g. in inter-bank communication). Obviously, as the number of remote organisations with whom one needs to communicate grows, this approach starts to become impractical!

Dial up

Either using PSTN and modems or ISDN—this can provide a low-cost approach for small user sites to connect to a number of different service providers (and is certainly the main method for members of the public to access service providers. ISDN can provide useful security through use of Calling Line Identification.

There are two main approaches to a dialled solution. First, the service provider could set up a number of modems (usually known as a 'hunt group') and the service user could dial directly to these. If use is on a large scale, it can be expensive and complicated for the service provider to provide sufficient modems, and also to address the need for wide geographic coverage of the service (so users can use local phone tariffs, rather than long distance rates).

A second approach is for the service supplier to use the services of a third-party network, with the network provider providing a large pool of modems (usually shared with other applications) with nodes in a large number of geographical applications. The service user will dial up to the network and will use protocols such as CCITT Triple X (X.3, X.28 and X.29) Packet Assembler Disassembler (in which case traffic is delivered to the service provider using high speed X.25 circuits) or if Internet, SLIP or PPP (the host will have a high speed fixed access into the Internet cloud via their service provider.

X.25

This technology developed with inter-organisational communications in mind, and today, remains one of the best technologies for the purpose. The X.25 standards developed when data communications were dominated by national PTTs and there were strong interests in ensuring national X.25 networks were inter-linked (using internetwork X.75 links and the global X.121 addressing scheme). If both organisations are attached to a public X.25 network, they should be able to communicate. X.25 connectivity provides useful security mechanisms such as closed user groups and Calling Address checking. One endearing feature of X.25 is that it tends to be a very economical option. On the down side, X.25 generally only suits the lower throughput applications.

Internet

Ever-increasing numbers of organisations are looking to this network for providing inter-organisational interconnectivity. It shares many of the advantages of X.25. Perhaps one of the biggest advantages over X.25 is that Internet is based on the TCP/IP suite. X.25 stops at the network layer of the protocol, and although sites may be able to interconnect using X.25, they may not be able to carry out meaningful applications communications, unless they have agreed common standards for interaction above the network layer. TCP/IP has standards running right up to application layer, increasing the likelihood that organisations can carry out useful interactions (e.g. file transfer, terminal log-on to applications, mail transfer etc.) without detailed investigation and possible development work.

Frame Relay

This is perhaps least suited to this role. Most networks are based on PVC connections, and do not allow on demand connections to remote sites. If two organisations are on the same supplier's network and they communicate frequently, in principle, they could have a PVC between them. Generally, FR networks are operated by today's global operators, and there has been less pressure to develop and deploy interconnection standards between networks—so if the parties are on different Frame Relay networks, there is little likelihood of interconnection today.

SMDS

This standard has many features that would facilitate high speed inter-organisational connections—a global addressing space and switched connectivity (in this case using connectionless protocols). Again at present, there is little interconnectivity between the SMDS networks of different suppliers.

A trivial approach—interconnectivity at the user site

In this approach, each user requiring inter-organisation connectivity will separately arrange this connectivity from their own site. This approach is the least good at handling aspects such as security and cost containment!

The approach works best where the user site already has enterprise network connectivity using 'traditional' networking techniques such as PSTN dial up or X.25 via a public network. In these cases it is only necessary for the user to establish a connection to the required organisation (indeed, with X.25 or a spare ISDN B channel, this could be achieved at the same time as maintaining the connection to the corporate enterprise network). Where the user has to obtain a separate network connection for inter-organisation connectivity, costs will clearly escalate if the practice becomes widespread.

A central network gateway

In this approach, the user traffic is carried to one of the hub sites over the corporate enterprise network. A router at this site would recognise inter-organisational traffic and forward it via a link to one of the possible inter-organisational network link types described above. It would be normal for such traffic to be focused through a gateway, so that the following functions can be carried out:

- Access security. Address screening is required to ensure that only those people permitted to communicate with the outside world do so, and only those in the outside world that are welcome have access to the corporate network. Routers performing this kind of security function are referred to as 'firewalls'.

- Address Translation. Commonly, a large enterprise network will have adopted its own private addressing scheme, that is inappropriate for wide area connectivity (an example might be an organisation that has chosen to use a private Class A IP address scheme, e.g. 10.x.x.x. This address range is completely ignored by Internet, while routers are configured to know that this address range is used for private networks). The company will generally solve this problem by owning a finite pool of legitimate public addresses (e.g. an IP Class C address range). These addresses will be assigned to legitimate inter-organisation users on a fixed or pooled basis. An address translation router is needed to map between a user's private network address, which is used for routing the traffic on the enterprise network, and the public address, which is used to represent that user of the enterprise network.

A central application gateway

This approach best addresses the issues of application compatibility. It assumes that an end user first communicates with an application running on a host/server within their organisation, and this software has the ability to communicate onward to hosts in other organisations, carrying out any necessary applications protocol translations that may be required. It is not uncommon for these application gateways to communicate with each other via an application enabling server in the wide area network. Typical examples of this are:

- E-mail systems—the user's mail will pass through a series of mailboxes, and inter-organisational mail will be delivered via a gateway (e.g. using a network supplier's X.400 system).

- EDI (Electronic Data Interchange)—this basically takes data records from one organisation's database (e.g. orders for goods or invoices) and delivers them to another organisation's database. To enable this, the data will be formatted in a standard format for its passage between organisations (e.g. EDIFACT). The end user will send the data via an application gateway in their enterprise host, which will perform the necessary translation. The data can then be passed to the recipient organisation, either directly, or commonly, via an EDI 'mail service' provided by the network operator.

5.10   CHECKLIST: AN 'A TO Z' OF TOP-LEVEL DESIGN ASPECTS

As this section has demonstrated, network design is a complex, multi-faceted task. This work is often done to demanding timescales. It is not uncommon for a network supplier to be asked to bid for such a network in a period of a little as four weeks from receipt of ITT to delivery of the tender! With such time pressures, it is all too easy for a significant, and perhaps costly aspect to be missed. Below is presented an A to Z checklist for the designer, to help insure that key areas have been properly addressed (It should be noted that this checklist will have to cover not only the physical aspects covered in this chapter, but also logical, management and core network aspects covered in later chapters):

❑ Accomodation. How much space is there for the communications kit—any special power or environment needs? Have costs for any racking been included?

❑ Addressing. Has the addressing scheme for the network been considered? What will need to be done to procure any public registered addresses needed?

❑ Assumptions and design issues. Have all assumptions made in the design and design issues been documented? Is action in hand to test and justify the assumptions and resolve the issues?

❑ Availability. Are components and services available now? What are the lead times? Are there any geographic restrictions?

❑ Cabling. Have costs for premises cabling been included—this can be a major cost element for the enterprise network. Cables from routers to end systems can also be a costly extra, especially if the router manufacturer uses an unusual, non standard connector, so that the cables can only be obtained from the manufacturer.

❑ Capacity. Do the site connections into the enterprise network have adequate bandwidth to carry required peak traffic loads? Have the core network nodes and trunk links sufficient capacity.

❑ Delay. Have calculations been performed to show that user transaction delays and file transfer throughput targets can be met?

❑ Growth. Have you considered planned growth? Also you will need to do 'what if' modelling for × 2 or higher.

❑ Implementation. With what we know from availability, are corporate plans for roll out likely to be met with this design?

❑ Infrastructure. Has the impact of this enterprise network on the core public network been fully assessed and costed? Will more network nodes and ports be required? Will more core bandwidth be needed? Are there any scalability issues with this growth (e.g. network may have to have a maximum number of nodes before it has to be split into different domains. There may be a maximum number of domains).

❑ Legacy. If there is an existing network and system, have methods been identified to support these legacy systems?

❑ Memory. Routers are often provided with only sufficient memory to perform the most basic tasks. If the router is doing anything other than the most basic TCP/IP routing, memory upgrades will be required. Further non volatile memory upgrades may be needed to facilitate software upgrades. Has the memory requirement been assessed and costed?

❑ Network management. Has an approach to management of the network been designed, and have the necessary components been included for costing?

❑ Originality. What if any aspects of the network design are original (in that they use new untested technology, or configurations of technology for which there is no known precedent?) What will be done to minimise risk?

❑ Protocols. Do we understand how all required user end system protocols are to be transported over the network?

❑ Repair time. Are we confident that means exist to repair the network in the required time? This will include costing of adequate support staff and spares holdings. Good back to back maintenance contracts with suppliers may be needed, which will have a price.

❑ Reliability. Have calculations been carried out to demonstrate that the network will deliver the level of reliability required?

❑ Resilience. Has sufficient fault tolerance been included in the design to meet the reliability requirement/ This is especially important at hub and host sites, where failure can affect very many users!

❑ Routing. What routing protocols are to run on the network?

❑ Security. Is the network sufficiently secure for the intended applications? Security includes the ability to deliver data end-to-end without loss, duplication or corruption, as well as the more obvious prevention of network access by 'hackers'.

❑ Software. This is a major cost element with routers, licences often cost more than the hardware. Check that the correct version has been specified, that supports all the features you wish to use.

❑ Statistics. Does the design include means to obtain traffic statistics, so that growth can be rationally planned?

❑ Staffing. Does the design consider the number of staff needed to operate, maintain and support the network? What will be the cost?

❑ Troubleshooting equipment. Handling of LAN/WAN faults often demands use of complex and expensive equipment and highly skilled staff, has this been allowed for? An alternative is to hire in consultants on a per occasion basis, who will come with the necessary equipment.

5.11   SUMMARY AND KEY ISSUES

Having completed the architectural design phase, the designer will have decided and documented the following:

- The number and definition of the site types to be used in the design.
- The technology and topology of LAN systems to be used.
- The topology and technology of the WAN system.
- The choice of gateway technology between WAN and LAN.
- How legacy systems will be handled by the gateway device.

There is a strong possibility that multiple design alternatives will come to light at this stage. In order to help choose the best, the designer should develop a series or 'thumbnail' designs using each of the contending solutions. The designs can then be compared using aspects such as:

- cost—outline costings can be done based on any known network supplier's tariffs for the WAN services, and list prices for LAN and gateway components
- performance
- likely reliability
- technical risk—is it proven technology or state-of-the-art?

By this stage in the process, we should now have a fully costed design ready to present to the customer. We have reached this stage by performing the following for each generic site type in the network:

- select LAN solution required at that site type;
- determine wide area network connectivity required;
- determine additional resilience networking requirements;

- select the gateway device (e.g. a router) needed to connect the LAN to the WAN.

In order to be confident that the proposed network will work as a whole and support the user's applications, we must also:

- Carry out sufficient logical design work to understand how the network will operate.

- Think through how the network will be managed, so that any additional management components can be costed.

- Carry out verification to confirm that the design will meet the customer's requirements.

The major issues associated with top-level design are:

- It is said that all useful networks fill to capacity, thus networks cannot be designed, they just evolve. Special care must be taken that the need for growth is anticipated at this design stage. Ensure that a way forward exists to increase connectivity and capacity of all network components. Further thoughts on future growth of the network will be found in a later chapter.

- It can be difficult to obtain good quantitative performance information from suppliers. Because routers are so complex, they may simply never have measured the performance in the configuration you plan to use. Where figures are provided, the exact test conditions may not be revealed, and of course, any figures quoted may be selected to show the product in a good light. Independent test houses often publish comparative test reports which can be of value (e.g. Kevin Tolly in Data Communications). At the end of the day, however, you may have to carry out some of these measurements yourself.

- Design is a compromise and an optimisation process—trading off cost, quality and functionality. The typical user wants a cheap network that is 100% available, is fast, and covers all current and future application needs. A certain amount of expectation management is called for on the part of the designer and this should be started as soon as possible.

- Customers may have an existing TDM/Digital Private Circuit network that supports both data and voice traffic—what happens to the voice traffic if a router + Frame Relay or router + SMDS data network is proposed? Future networks using ATM or IP VPN technology should be able to carry both voice and data. Voice can be sent over the Frame Relay networks of today, but this requires sophisticated voice compression techniques together with devices capable of accommodating the variable delay encountered in these networks. Frequently, a separate voice only network will be more cost-effective, provide better voice quality and a greater range of functionality.

- 'Cloud mindedness'—it is all to easy to fall into the trap of portraying the wide area network as a nice friendly fluffy cloud; this can be a useful abstraction at times! However, the designer must carefully think through the impact of the design on the cloud and of the cloud on the design. Examples are:

  - If the proposed network is large, will the network provided have to expand (in terms of the number of ports and bandwidth between nodes) to cope? If so, how long will this take? Is the underlying network technology suitably scalable?
  - The network will invariably be implemented as a partial mesh. If the customer has a major hub site, what is the best way to connect this site into the cloud? It may be necessary to increase the amount of meshing bandwidth and paths to conduct the necessary traffic levels through the network to the nodes where the host connects. It may be more cost-effective to connect the host site via longer length connections to those network nodes with the largest numbers of branches attached—so the majority of traffic does not have to pass over meshing paths at all!
  - What really happens when a network node or trunk link fails? How long does it take to re-establish connections via an alternate path—how good will the alternate path be? It may be congested or be much longer, adding significant delay.

BIBLIOGRAPHY

Books that cover the key aspects of high-level design are few and far between. The titles we have found to be useful are:

Perlman, R. (1992) *Interconnection—Bridges and Routers*. Addison-Wesley.
Atkins, J. and Norris, M. (1998) *Total Area Networking*. John Wiley & Sons.
Spragins, J. (1991) *Telecommunications Networks, Protocols and Design*. Addison-Wesley.
Held, G. (1998) *Internetworking LANs and WANs*. John Wiley & Sons.

6

Logical Design

Opportunity is missed by most people because it is dressed in overalls and looks like work

Thomas Edison

In the previous chapter we saw how to design the enterprise network in terms of the physical components that make it up—the LANs, the routers and the wide area network connections. In this section we address the logical aspects of the design. This covers how we configure the components and how they then interact as a whole system to support the users and their end-to-end applications. The key aspects that we shall examine are the setting up of addressing schemes, the choice and configuration of routing protocols, the inclusion of appropriate security mechanisms and the support of end system protocols (including the support of legacy systems).

Our purpose in this chapter is to examine the key design issues surrounding protocols, rather than the technical aspects of the protocols themselves. There are many excellent texts that explain the technology available to the designer. Readers who wish to gain a more in-depth knowledge of these are referred some of the books listed at the end of this chapter (in particular: Comer 1995; Stevens 1994.)

6.1 NAMING AND ADDRESSING SCHEMES

Many 'traditional' data networks (e.g. point-to-point private circuits and Time Division Multiplex networks) operate at the physical layer of the OSI seven layer model, and provide simple 'bit pipes' between communicating end systems. Once the network supplier has configured the bit pipe, the associated end systems have the potential to communicate over it.

The advent of packet oriented switched networks (e.g. X.25, TCP/IP networks and LAN technologies) changed this picture. In these networks, data is sent in blocks referred to as packets. Network switching decisions are made on a packet by packet basis, operating at the network layer of the OSI model. To function correctly, a coherent addressing scheme must be set up, so that a network can make appropriate routing decisions.

One of the first tasks of the network designer at the logical design stage is to design the addressing scheme for the network. The exact requirements of the addressing scheme will depend on the protocols to be operated on the enterprise network (and of course, where this supports multiple protocols, the addressing scheme for each will need to be designed). The addressing scheme will be designed with the following in mind:

- The need to identify uniquely each separate end system. In any viable enterprise network it is vital that each communicating device (PC, server etc.) can be reached and this implies an overall address scheme.

- The need to reflect the topology of the network (generally this will consist of a set of 'subnetworks' interlinked by switching nodes or routers). Network switching nodes use the topological information, inherent in the addressing scheme, to build up 'route maps' allowing them to make the necessary routing decisions.

Once the addressing scheme has been designed and documented at an abstract level, there is a need for a network administrator to issue, record and maintain the actual addresses used. Failure of this role can have serious consequences for the correct operation of the network, especially in the light of the agility required to support the modern enterprise. (An illustration of this is given at the end of this section.) Some examples of protocols where addresses have to be allocated are now given, together with some notes on designing the addressing scheme for that protocol:

- LAN Media Access Control (MAC) address. This is the link layer address of individual LAN interfaces in the end systems and routers. Often, these addresses are hardwired into the interface card—standards exist so that each card manufactured will have a unique address. This is not always the case however, and some network systems require MAC addresses to be allocated and configured into systems (e.g. this is common in IBM SNA systems).

- Token Ring. Token Ring networks (especially when supporting IBM SNA applications) tend to make extensive use of Source Route Bridging technology to allow packets to be correctly conveyed across multiple LAN segments interconnected by bridges. In these networks, it is essential that the network administrator assigns a unique ring number to each Token Ring. Numbers are also assigned to bridges, but the only time a bridge needs a number other than a default value of 1 is if there is more than one bridge directly interconnecting two rings (as might be done for added resilience against bridge failure).

- IP. It seems inevitable that at least some entities in an enterprise network will need IP addresses, even if the network uses bridging technology and no end-to-end applications use TCP/IP stacks. The reason for this is the ubiquitous use of SNMP management for networking elements—and this protocol relies on IP for its communications with network management systems.

IP addresses are 32 bits in length (conventionally, addresses are written as the decimal values of the 4 bytes, separated by dots. By way of example, a binary address 10000001000000010000001000000011 would be written 129.1.2.3. The address is divided into the network address and host address as in Table 6.1.

Table 6.1

CLASS (MSBs)	Network address length (bits)	Host address length (bits)	Number of networks	Number of hosts
A (0)	7	24	126	16 777 215
B (10)	14	16	16 382	65 534
C (110)	21	8	2 097 150	254

Note: The most significant bits (MSBs) of the address are used to distinguish which addressing class is being used. In addition to these three addressing classes, the designer may also need to work with class D addresses (MSB 1110). These are used specifically for designating multicast groups (e.g. used for some routing protocols or for voice/video distribution applications).

When designing an IP addressing scheme, the administrator's first job is to decide if the network should use a publicly registered address or a private addressing scheme. Publicly registered addresses are normally adopted where the user has a need to be connected to the public Internet. The administrator will have to apply to the Internet authorities (InterNIC or RIPE) for a registered addressing domain (the class requested will depend on the size of the network, in terms of the number of hosts). Note that due to the rapid expansion of the Internet, IP addresses are in short supply. Class C addresses are normally available but class B are often difficult to obtain (a good case is required for one to be allocated) and registered class A should be considered, to all intents and purposes, unobtainable.

The shortage of registered IP addressing space has been recognised, and the IETF (the guiding body for the technical development of the Internet) have developed a next generation IP protocol, IPv6 (today's version is IPv4). This new scheme has, amongst other features, a much longer addressing field that allows many more addresses. However IPv6 is not widely implemented at present.

Where there is no need to interconnect to the Internet, the user can create their own address scheme. It is recommended that the administrator should follow the guidance of RFC 1597 and adopt the class A address 10.x.y.z. (There is also a reserved range of class B and C addresses for unregistered private network use). The public Internet is set up to discard any packets received bearing an address from this range.

The most common case will be a large company that *does* need to connect to the Internet (to provide its own employees access to the Web or e-mail, or to let is customers and trading partners access to Web servers or other applications provided by the company). Because of the security risks associated with Internet connection, this should always be done in a controlled manner, through a company firewall. The fact that the interconnection is done in that way can be used to advantage from the addressing point of view. The company will use an unregistered addressing scheme internally. It will then obtain a registered address range for Internet facing activity. The firewall will carry out network address translation, mapping between the internal and external addresses as necessary. Network address translation is further discussed in section 6.2

At this stage, the addressing scheme defines a single network with many hosts. In reality, an enterprise network will need to divide this space up into many 'subnetworks', each with its own subnetwork address. Subnetworks are interconnected by routers, which will not forward traffic intended for the local subnetwork, and will know the best path to forward traffic to remote subnetworks. The network administrator will need to decide how to split the host address space to define a range of subnetwork addresses. This is shown in Figure 6.1.

In the figure, it has been decided to use the RFC 1597 10.x.y.z range. The network administrator has decided that there will never be more than 254[1] hosts in a subnetwork. Therefore, the least significant 8 bits are the host

---

[1] The reader may be wondering why the number 254 is used, when 8 bits represent 256 possible values. The reason for this is that IP reserves 255 as broadcast address. When this value is used in a destination address, all devices on the subnetwork will receive and process the packet. A value of 0 also has a special meaning, referring to 'this subnetwork' in configurations.

Network Address	Subnetwork Address	Host Address
10.	X.Y.	Z

Figure 6.1 Subnetwork Addressing in IP

addresses and the 16 bits above that can be allocated to support up to 65534 subnetworks.[2] This is indicated, in the end system and router configurations, by a subnetwork mask. This has binary 0s representing those bits that are reserved for host addresses (it is known that class A has 8 bits of network address, B has 16 and C has 24, so the number of bits in the subnetwork address file can be inferred). In the Figure 6.1, the mask is 255.255.255.0. When quoting an IP address or configuring it into a device, it is generally necessary to give the mask as well as the address. Figure 6.2 shows this addressing scheme applied to an example network with 4 subnetworks—three user sites and a wide area network.

In the diagram, site 1 has subnetwork address 10.0.1.x, with hosts allocated IP addresses 10.0.1.2 and 10.0.1.3. the router has the address 10.0.1.1 allocated to its LAN port. A similar pattern of subnetwork address and host address assignment is applied to sites 2 and 3, together with the WAN subnetwork.

Further considerations in designing the address scheme

Of course, the designer is under no obligation to stick with an 8-bit boundary for subnet masks. If we decide that we need 1022 hosts for each subnet we can use a mask of 255.255.252.0. Similarly, if 30 hosts is enough, we can use a mask of 255.255.255.224. In practice, it is always best to err on the side of caution, because if a subnet runs out of host addresses, this can force you into renumbering the whole network and this can be a painful and time consuming exercise. In a network, one can have the same mask for each subnet—this is certainly the simplest approach and allows use of the simpler routing protocols. This can however, be inefficient where there is a mix of site types of differing sizes. Variable length subnetting is an alternative approach, but a more complex routing algorithm that can handle this is required.

In assigning subnetwork numbers, it can often help network operators if a meaningful number is chosen to represent a particular branch network,

---

[2] Again, values of 0.0 and 255.255 are reserved for 'this network' and network wide broadcast.

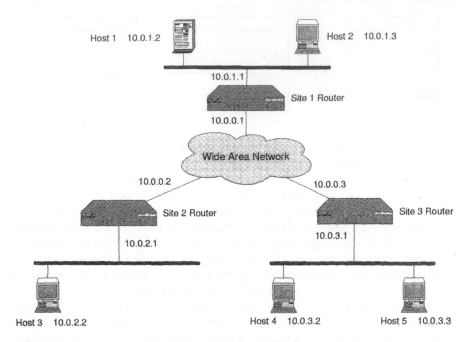

Figure 6.2 Addressing Schemes Applied

e.g. if branches already have numbers for company administrative purposes, this number could be selected. If it is an international network, a country code (e.g. international phone code) could be used.

It can also help the network operators if certain addresses are used to indicate specific types of subnetwork, e.g. in Figure 6.2, 10.0.0.x is reserved for WAN links. This particular figure shows the very specific case where a single subnetwork can be assigned for the wide area. This is only the case if the router can communicate with all other sites via a single access port. This is the case for multicast networks such as SMDS and with networks such as Frame Relay. These types of network are often referred to as Non Broadcast Multiple Access (NBMA networks). If the WAN consists of a set of point-to-point connections, each connection must have a subnetwork address—one for each port on the router. As there are only two 'hosts' (each end of the link), this can be wasteful of address space. A solution here is to use a different mask—say 255.255.255.252. If this is done, a routing protocol (see later) will have to be chosen that is capable of supporting variable length subnetwork masks. The OSPF protocol is the prime candidate for this as the simple (but ubiquitous) RIP protocol does not support variable length subnetwork masks, and the user would be forced to allocate a full 254 host subnet for each circuit if it is used.

Mobile IP users

An increasing problem for network designers is the rise of flexible working practices. Many users today use laptop PCs exclusively. The user reasonably expects to be able to plug this in to any LAN socket, at any desk in any company building, to be able to access e-mail, the corporate intranet and other applications required for the job.

It is impractical for a network administrator to hand out a different IP address for each building (or often for each floor or office within a building) to the user. Its also an inconvenience for the user to work out which IP address to use in any location and to configure the PC to use it. The answer to this problem is to use a protocol called DHCP, which allows the dynamic allocation of IP addresses. To use this, the designer must provide one or more DHCP servers in the network. These keep a pool of free addresses for each subnet. Users' end systems are set up to send a broadcast message when they start up to locate the DHCP server. This will allocate the user an appropriate free IP address, which the user's PC then uses. DCHP is not just good for laptops, it can be used for desktop PCs as well thus saving on the administrative overhead of managing IP addresses for these.

Novell IPX

This is one of the most widely used protocols for office automation networks, and many enterprise networks will have to support Novell traffic. Addresses consist of two parts, the network address and the host address. Unlike IP, where the host address has to be assigned, IPX uses the station's MAC address as the host address. The network administrator has to assign a unique network address to each subnetwork. This is a common cause of difficulty when attempts are made to interconnect previously separate Novell networks, as often the network address on the separate networks has been left as the system default value of 1. Significant effort may be required to reconfigure the different networks with unique addresses. To assist with the assignment of unique addresses, Novell now offer a network address registry, to which users may apply for unique Novell network addresses.

An example of failed network administration

Before leaving the topic of addressing schemes, let us look at an example (based on actual experience) of what can go wrong if careful network administration of the address space is not followed. Figure 6.3 illustrates this.

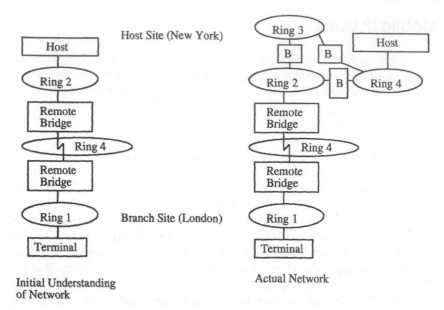

Figure 6.3 An Example of Failed Network Administration

The story is this: a small airline has its main office in New York, and a branch office in London. The host is an IBM AS400 which is Token Ring attached. The branch has PC terminals running an IBM terminal emulation package, and these attach to a Token Ring LAN. The airline approached a supplier of managed networks to interconnect the sites so that terminals in London can use airline reservation applications running on the New York AS400. This was achieved using Remote Source Route Bridging. In fact, routers were provided at each site and these were interconnected using Frame Relay. Cisco proprietary Remote Source Route Bridging was used, with the two sites interlinked using TCP/IP sessions, which carry securely the Token Ring frames. This technology works by simulating a central 'virtual ring', and each router connecting to the virtual ring simulates a bridge. Now, the connectivity was ordered by the London office, and they informed the managed network supplier that the London LAN was Ring 1 and the New York router and Host would be on Ring 2. The customer allocated the network supplier Ring 4 for the virtual ring.

When tested, the terminals in London could not connect to the New York host. Investigation revealed that the New York environment was now a complex multi ring environment, consisting of three rings interconnected by bridges. The New York router was attached to ring 2 as expected, but the target host was on another ring, reached by a bridge—and that ring was numbered 4, the same as the virtual ring!

In Source Route Bridging, a terminal finds its host by sending out a broadcast explorer packet, which is copied by bridges and passed on to other rings attached to the bridge. As the explorer passes through a bridge, the bridge adds details of the ring number that the packet came from, together with the bridge number, into a Routing Information Field (RIF) in the explorer. When a copy of the explorer finally reaches the host, the RIF will describe a complete route back to the terminal, which can be used in all further communications between terminal and host. In order to prevent needless propagation of explorers through the network, the bridges will not send an explorer into a ring where it has already been—therefore explorers from the terminal never reach host ring 4—as the RIF field of these explorers already includes a ring 1 (i.e. the Virtual Ring) when they arrive at bridges leading to the host ring.

This problem was easily resolved when the network administrator assigned a new, unique ring number for the virtual ring.

6.2 NETWORK ADDRESS TRANSLATION

If the network designer has followed the advice to use the public unregistered addressing ranges in the enterprise network, a number of problems can subsequently be encountered:

- What if the company wants to have an Internet connection, to allow its employees access to the World Wide Web and the ability to do Internet e-mail? The company will have to use registered addresses for this.

- What if the company wants to offer information about itself or provide goods and services via the Internet? Again, registered addresses will be needed.

- What if the company takes over or merges with another company, whose network also uses the same unregistered address scheme? There would be great difficulty in interconnecting systems if the same addresses could be found on both networks!

The answer to these problems lies in the use of network address translation. This function can be carried out by routers, or most commonly by a firewall device. A typical set-up for NAT is shown in Figure 6.4:

Suppose the company has decided to use the 10.0.0.0 unregistered class A address range internally. The company wishes to publish information about itself and its products on the World Wide Web, and places this material on a server with address 10.0.0.1. (shown in the top half of the diagram). Now, this address would not be recognised by the Internet, which is configured to ignore the private address ranges. To solve this problem, the company has obtained a registered class C address range, 193.14.71.0, mask 255.255.255.0. They have allocated address 193.14.71.5 to represent the Web server. Let us assume there is a remote user with

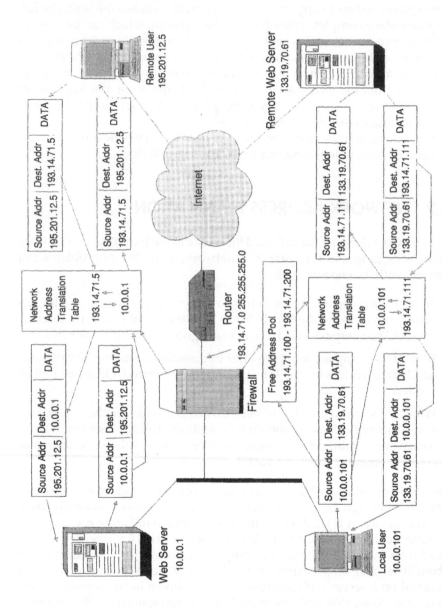

Figure 6.4 A Typical Set-up for Network Address Translation

address 195.201.12.5 that wants to look at the company's Web site. The remote use will send an IP packet to the Web server to request a page. This contains the remote user's address (source address) and the registered address of the Web server. This packet is intercepted by the company firewall, which has been programmed with a static address mapping for the Web server.

The firewall will modify the destination address field of the IP packet, replacing the registered server address with the private local address, before forwarding the packet into the enterprise network. When the server responds with the required page, the IP packets carrying the data will again be modified by the firewall, so that the remote user perceives the response is coming from the registered source address and not the company's private address.

The lower half of the diagram looks at the case of a company employee, using a PC with address 10.0.0.101, wanting to browse a remote Web server—in this case with address 133.19.70.61. To allow outgoing Web access, the network designer must allocate a pool of free addresses from the registered address range. This pool is configured into the firewall. When the local user sends an IP packet toward an Internet address, it is intercepted by the firewall. It will look up the source address in its address translation table. If there is no entry, it is assumed a new session is being started, and a free address from the pool will be allocated to the user. In this case, our user, 10.0.0.101 has been allocated free address 193.14.71.111, and an appropriate mapping has been placed into the table.

The firewall replaces the private source address with the registered address and forwards the packet to the Internet. When the remote server responds, its IP packets will be set to the registered destination address, so they can be correctly routed through the Internet to the company firewall. Here the address translation process replaces the destination address with the user's private address before forwarding responses to the user. A key design task here will be to ensure that there are sufficient addresses in the free address pool to ensure that local users are likely to find one free when they come to use the Internet! One small detail that also needs to be checked is that unregistered IP addresses do not escape in the payload of the packet.

6.3 NAMING

Many networking protocols allow users to refer to client and server hosts using names rather than numeric addresses—this makes life much easier for the end user! As much care needs to be taken in designing naming schemes as with addressing schemes. Each networking protocol has its own naming requirements. Two common examples are introduced below:

TCP/IP

Anyone who has used the Internet World Wide Web will be very familiar with using host names. To read a page of information, you select 'http://domain.name/pagename'. Similarly, when you send an e-mail, it is addressed to 'a.person@domain.name'. In both these cases, domain.name represents a specific server, and the name has to be translated to an IP address. In the simplest case, this can be done using host tables contained in the client end system. It is much more common however to set up a network server called a Domain Name Server (DNS). Client systems are configured with the IP address of the DNS, and they use the IP name resolution protocol to ask the DNS to translate a name into an IP address (or vice versa).

In large networks, we need a resilient design for the DNS. It must also be capable of coping with the rapid amount of change that exists in a typical network—with new server names appearing (almost literally) daily. This is achieved through a hierarchical structure of DNS servers, set up in such a way as to reliably propagate changes through the network. Client systems are set up with a list of possible DNS servers, allowing them to resolve names via alternative resources if their preferred (normally their nearest) DNS server fails. DNS design is a complex area, and the designer will need to refer to a specialist text for more information on this (Norris and Winton 1997).

If a company needs a presence on the Internet, it will need to obtain a formally registered domain name. This is usually arranged by the company's Internet Service Provider (ISP), which provides a point of reference to the public Internet complete with a standard range of service features (Frost and Norris 1997).

NetBios (or NetBEUI)

This widely used protocol is unusual in that it does not use network addressing at all—indeed the protocol has no network layer, being designed for use exclusively on bridged networks. Instead, each client or server on the network must be allocated a unique name. Careful network administration is needed to ensure the names remain unique. To connect to a named destination, a user's PC will send out a name resolution broadcast, which will be received by every device on the network. The owner of the name will respond, allowing the user to discover the LAN MAC address of the required device.

NetBios is somewhat of a legacy protocol but is widely used by many user applications. Its bridged nature makes life difficult on large networks, which, due to their size, tend to be router based. Today it is quite common to find NetBios packets actually transported over TCP/IP sessions, and this approach is widely used in Windows NT based networks. In these networks, we need to be able to translate a NetBios name into an IP address. This is done by providing a Microsoft WINS server on the network.

As well as designing naming schemes for network protocols, there are other areas where the network designer and administrator need to do work designing naming schemes. An example is e-mail addresses. A method must be found to ensure that each company employee is assigned a unique e-mail mail box name.

6.4 ROUTING

Having set up the addressing scheme, we need to decide how routers in the network will acquire the necessary information to route the packets of user information across the network. This is achieved by selecting and configuring an appropriate routing protocol. These protocols are responsible for exchanging routing information between routers, and helping each router to build a routing table (this will show, for each possible destination subnetwork, which router the user's data packet must be sent to next, in order to reach that destination).

A range of different routing protocols exist, each with their own strengths and weaknesses. To further complicate matters, in a multiprotocol environment, each LAN protocol being routed may need to use its own routing protocol to exchange addressing information specific to that protocol! Selecting the optimum routing protocol (or mix of protocols) is one of the key decisions of the logical design process. Below, a number of routing protocol options are described to illustrate the range of possible choices:

Static routing

In this somewhat trivial approach, routing is achieved through route statements explicitly encoded into the configuration tables of each router. If we take the example shown in Figure 6.5 and consider traffic flowing between LANs 1 and 2, then Router 1 would need a routing statement showing that subnetwork 10.0.2.x was to be reached by sending packets via its WAN port on subnetwork 10.0.1.x. Similarly, for return traffic, Router 2 will need a configuration command stating that subnet 10.0.0.x can be reached via its WAN port on subnetwork 10.0.1.x. Static routing cannot make effective use of resilient pathways—such as via router 2—so configuration to route traffic via this route would not normally be added to routers 1 and 3.

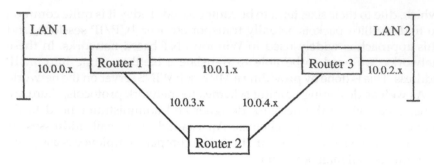

Figure 6.5 Optimum Network Path and Resilience

Static routing has the advantage of being relatively simple to configure in small networks, and requires little processing power from the routers to operate. It can also be invaluable where security as at a premium, as one has exact control over how traffic flows through a network. There is no opportunity for dynamic routing algorithms to discover unforeseen insecure paths or for systems on interconnected public networks to learn about the statically configured subnetworks.

While static routing has some advantages, the disadvantages are so great as to relegate the use of this form of routing to special circumstances only. Disadvantages include:

- Lack of scalability—as any new subnetwork is added, all routers that need to communicate with it, or are involved in switching traffic for that network will require to be reconfigured with details of that subnetwork. With networks of more than a dozen routers, this task starts to become impractical.

- No resilience—a key advantage of dynamic routing protocols is that networks can self heal by discovering alternative paths to destination subnetworks, where these have been provided. Statically routed networks cannot take advantage of alternative routes in this way.

Dynamic routing—distance vector protocols

The RIP (Routing Information Protocol) from the TCP/IP suite is a classic example of this type of protocol. Novell networks also use a proprietary variant of RIP for routing. Routers running this type of protocol are just configured with a list of the subnetworks directly attached to that router. The router will periodically broadcast a routing update to all other routers attached to these subnetworks, telling them about the subnetworks that can be reached. When a router receives an routing update from a neighbour, the update will contain information about subnetworks not directly connected to the recipient router. The router will merge this newly acquired information into the routing table. When this router next sends its own update, it will pass on information about the networks it has just learned. This process is repeated until eventually, all nodes have learned routes to all subnetworks.

From the above description, two of the disadvantages of dynamic routing protocols are clear. First, they take up network bandwidth as they send their routing updates and second, they require processing time in routers to send and receive these updates, and carry out associated routing table maintenance.

Achieving optimum network path and resilience

So far, we have not explained why this type of routing protocol is called a Distance Vector Protocol! This is best explained by an example. Consider the network shown in Figure 6.5. As well as the subnetwork addresses that can be reached, the routing update contains information about the 'distance' to the subnetwork. With RIP, this is simply expressed as the number of hops (i.e. the number of routers that must be passed through to reach the destination subnetwork).[3] So in Table 6.2, updating would proceed as follows for router 1 (the other routers will operate in a similar fashion).

The first thing to note is that in the initial condition (e.g. after router power up, router 1 does not know how to get to subnet 10.0.2.x. It takes one update period to get this information. For every extra hop a destination is away, it takes another update period to learn about it. The time taken for a router to fully learn about all available subnetworks is called the convergence time and can obviously be as high as 15 update periods. The convergence time is an important design consideration, as it sets an upper limit to the time it takes the network to recover from a failure (following which the routing tables may have to be relearned). In the design one can trade-off the update period (the faster this is the more bandwidth and router processing power required) with convergence time (the more frequent the update, the faster the network can repair itself).

After two update periods, router 1 learns of a second route to 10.0.2.x, via router 2. This route has a hop count of 2 – higher than the previously learnt route. This longer path is only ever used to send traffic if the shorter route fails and in subsequent routing updates the shorter route is removed from the routing table.

---

[3] An important limitation of RIP that network designers must consider is that the maximum distance possible is 15 hops. Any subnetwork 16 hops or more away is unreachable using RIP.

Table 6.2

Update	Subnets known	Hop count
Initial condition	Locally attached subnetworks only	
	10.0.0.x	0
	10.0.1.x	0
	10.0.3.x	0
After update 1	Locally attached plus subnets learned	
from routers 2 and 3	from update	
	10.0.0.x	0
	10.0.1.x	0
	10.0.3.x	0
	10.0.2.x	1 (via 10.0.1.x)
	10.0.4.x	1 (via 10.0.3.x)
After update 2	Locally attached plus subnets learned	
from routers 2 and 3	from update	
	10.0.0.x	0
	10.0.1.x	0
	10.0.3.x	0
	10.0.2.x	1 (via 10.0.1.x)
	10.0.4.x	1 (via 10.0.3.x)
	10.0.2.x	2 (via 10.0.3.x)

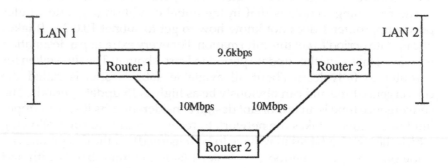

Figure 6.6   A Limitation of RIP

More sophisticated distance vector protocols

RIP is a very simple routing protocol, able only to use hop count in determining the best path through a network. Consider the scenario shown in Figure 6.6.

In this rather extreme example, router 1 and 3 are linked by a 9.6 kbps serial link. The routers are also interconnected via router 2, where router 1 is linked to router 2 via a dedicated 10 Mbps Ethernet segment, and router 2 links to router 3 via a similar segment. Relying solely on hop count, RIP would choose to send traffic from LAN 1 to LAN 2 via the 9.6 kbps link! (In reality, a designer faced with this situation could manually configure routers 1 and 3 to add a hop count offset to any routes learned over the 9.6 kbps link, forcing traffic via the 2 hop fast route under non failure conditions. Of course, such hand crafting is undesirable and becomes increasingly difficult to manage as networks become larger.)

Several proprietary variants of the Distance Vector Protocol exist which overcome this limitation (these include the Novell RIP and the Cisco IGRP protocol). Both replace simple hop count by associating a 'cost' to each network link. This metric is then sent along with any routes broadcast over that link. The costs are calculated by proprietary algorithms from a number of factors. These include bandwidth, round trip delay and reliability. The routing algorithm selects a path to the destination, based on the lowest overall cost, rather than just hop count.

Dynamic routing—link state protocols

The Distance Vector Protocols suffer from two main disadvantages—they use a lot of bandwidth continually sending out routing updates, and they can take several minutes to reconverge if a link fails. (As the change ripples through from one router to the next only at each update time, note that proprietary variants are able to do forced updates when changed routing information is received, which speeds reconvergence.)

To resolve these difficulties, a new generation of routing protocols has been created called the Link State Protocols. The TCP/IP version is known as Open Shortest Path First (OSPF) and the Novell version is NLSP. During normal working routers simply exchange 'hello' messages with their neighbours, to check that the links are working OK. These messages are short and use little bandwidth. Only if a hello handshake fails do the routers then enter a recalculation process, where all routers will recalculate the network topology.

When the network first starts up, routers will broadcast a message out of all its interfaces to try to discover its neighbours. Once a router has done this, it will broadcast a neighbour list to all routers in the network. A router is now able to calculate a complete network map from the all the neighbour lists received. Only a failed link (or a new link added to the network) will now trigger a recalculation process.

The downside to the Link State Protocols is that the recalculation process is complex and powerful router processors and much memory are needed. Even so, it can take many seconds to complete—during which time the network can pass no traffic. That said, reconvergence after network failure is generally much quicker than Distance Vector, making Link State algorithms the method of choice in many networks today.

This section can only offer a superficial view of the OSPF protocol. In reality it is a complex protocol and will require specialist knowledge to design stable networks using it.

Scalability of link state protocols

With the Link State protocols, the time taken to recalculate the network map increases as a function of the number of routers in the network. This in principle, should make scaling such networks a problem. The reality is that Link State Protocols were designed with scalability in mind—and this is done by dividing the network into areas. An area is a set of routers sharing in the same map recalculation process—typically one would have up to 50 routers in one area. Areas are linked by boundary routers (which typically are connected to a maximum of three areas). Boundary routers act to decouple the recalculation process. A typical multiple area topology is shown in Figure 6.7.

In the figure, there is a central backbone area 0. This is a set of interlinked routers, the links being portrayed by the central 'cloud'. Area 1 is also a set of routers linked by its cloud. Areas 0 and 1 share in common two boundary routers (one boundary router between areas is sufficient, but large networks often call for two to eliminate a single point of failure). Area 2 is similarly linked to the core. In a typical hub and branch network, hosts and hub servers are generally associated with routers in the backbone area. Any peer to peer traffic (say between Area 1 and 3, would have to pass via the backbone area.

Using this multiple area technique, networks can, in principle, be scaled to several thousand sites. In reality it is safer to set design limits closer to 500 routers in an OSPF design.

Routing summarisation

In the network shown in Figure 6.7, we can significantly increase the efficiency of the routing process if routing summarisation is used. This works best with a carefully chosen addressing scheme which reflects the area structure of the routing protocol. If we keep to the 10.x.y.z addressing scheme, let x = area number, y = subnetwork and z = host. In this case, the boundary routers linking areas to the backbone can 'summarise' routing information. They only need to tell other backbone routers that they can see 10.x, rather than provide a full list of all the subnets 10.x.y.

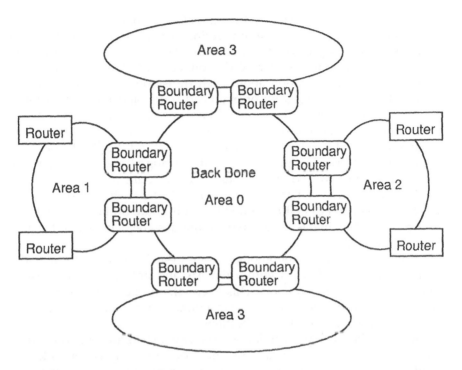

Figure 6.7 Typical Multiple Area Network

Multiple autonomous systems

In the sections above, it was recommended that OSPF networks should not typically exceed 500 routers—although in well structured hierarchical designs the limit can approach 2000. What happens if we need larger networks? Indeed what happens if we need to interconnect the network of many companies? The answer lies in the use of multiple autonomous systems (referred to as AS).

An OSPF network consisting of a backbone and petal areas represents a single AS. The OSPF routing protocol itself is described as an interior gateway routing protocol. Autonomous systems can be interconnected using an exterior gateway protocol which would run on a router attached to each OSPF backbone network. These AS gateway routers are interconnected via a backbone network. For the sake of network resilience, one would generally have at least two such gateways for each AS. The main exterior gateway protocol in use today is BGP4. (BGP can be applied to both exterior (eBGP) and interior (iBGP) gateways. Here we assume the former.) Is this scalable? Well, this is the technique used to build the public Internet!

BGP4 is designed to communicate only highly summarised information about the networks attached to the AS gateway router. Because of this, BGP4 is very effective at containing routing problems within autonomous systems. Suppose a given subnet within an OSPF area kept appearing and disappearing—due to a faulty WAN connection (often referred to as a 'flapping link'). Information about each transition would be circulated around the AS, creating a lot of overhead traffic and requiring a lot of router processing power. Because BGP4 only passes information about the reachability of whole networks, the overhead caused by the flapping link is not propagated to other networks. As a result of this, BGP4 can create very stable networks. Designers faced with creating a large single enterprise network are often well advised to think about creating multiple AS linked by BGP4 for this reason. The downside of BGP4 is that network convergence time is not as good as OSPF, finding the right trade-off between stability and convergence time is always part of the designer's job when tackling routing protocols!

Ideally, one designs autonomous systems in such a way that the majority of traffic will be delivered within the AS, and only a minority crosses between autonomous systems. This keeps the number of routers traffic has to transit (hence network delay) to a minimum and also helps prevent the gateway routers becoming a bottleneck.

Figure 6.8 shows a typical multiple AS network. There are four universities, each with their own large campus network. The campus networks use OSPF routing. The universities have decided to interconnect their networks to allow sharing of information, e-mail etc. This is done by a Universities Backbone Network, which runs BGP4.

Centralised routing—the route server

So far, we have discussed distributed dynamic routing protocols. Another approach to routing is to use a centralised routing server. This method has been used on many connection-oriented networks. Each router will have a permanent path established to a central server. When the router has a packet to send to a destination with which it has not communicated before, it will apply to the central route server for a route to that destination. Once received, the router will forward that packet on the specified route, caching the route to be used for any subsequent packets to that destination.

This approach is now widely used with the introduction of ATM LAN emulation. The technology provides high speed LAN switching (i.e. bridged) interconnectivity between sites (and there are various standards, such as LANE and MPOA). LAN switches in the network can find the location of a given MAC address by referring to a central server. This returns the ATM network address of the remote LAN switch handling that MAC address, enabling an ATM call to be established to that switch.

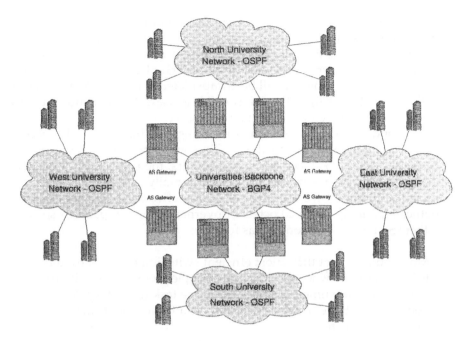

Figure 6.8   A Typical Multiple AS Network

Service advertisement protocols

Closely related to routing protocols are Service Advertisement Protocols. These are used in client–server systems, where the available servers periodically advertise their services to possible clients. SAP protocols are most likely to be encountered by those working with Novell, where the servers will broadcast their presence every 60 seconds. The SAP broadcasts will normally be propagated throughout the network by routers, and in a large network this can create a significant traffic overhead. When designing with Novell networks, therefore, an important aspect will be to decide on the minimum set of services that each site needs to advertise, and the design of a set of filters for the routers, which will prevent unnecessary SAP traffic being sent into the network. It may also be necessary to configure routers to send SAP updates less frequently, to reduce network overhead.

6.5   PROTOCOL SUPPORT

When it comes to supporting modern LAN protocols, routers can use two principal methods. The first method is native transport. In this mode, all routers in the network will run the routing algorithm designed as part of that protocol suite. The protocol data packets will be carried 'as is', encapsulated efficiently onto the bearer protocol of any network used to interconnect the routers (be it Frame Relay, SMDS or HDLC for point-to-point links). Ideally, the protocol will be encapsulated using a standards based method (e.g. RFC 1490), so that it is possible to use different manufacturer's equipment at either end of the link if necessary.

A second approach is that of tunnelling. In this case, only 'access routers' directly attached to the traffic sources would run the native routing protocols. Between the access routers, the protocol data packets are encapsulated in a bearer protocol (normally IP), and the traffic is treated as though it were IP over the core network. To do this can look inefficient, adding up to 20 octets of overhead to every packet—so why bother? Some of the reasons are as follows:

- Most routers are at their most efficient switching IP. Also, routers tend to be less efficient when dealing with multiple protocols, as their processor will spend more time switching from one task to another. Many large networks are therefore designed to have a fast efficient single protocol (normally IP) core.

- Where there is a minority protocol used at only a few sites, we can avoid the need to configure too many routers with this protocol—only those at the affected sites need to be set up with the encapsulation.

- Non TCP/IP protocols may be forced to use simple routing algorithms that are slow to reconverge. If this traffic is encapsulated and delivered to its destination by a TCP/IP network, that protocol will enjoy the same fault recovery properties as the TCP/IP network. This can be especially important where a site is dependent on both TCP/IP and another protocol such as Novell to perform the business applications required at that site.

6.6 NETWORK SECURITY

All data networks are a potential target for attack by the malicious, or simply curious, 'hackers' and, increasingly, network attacks are motivated by criminal intent. This could be as simple as someone stealing personal data and selling it (e.g. to detective agencies) or it could be a direct attack on a bank's records.

Security is a serious issue that the designer cannot afford to overlook. Major security incidents as recorded by the CERT emergency response team are currently running at some 2500 incidents a year. There may well be many more that pass unnoticed.

Network security is all about safeguarding the operation and preserving integrity in the face of accidental damage or deliberate attack. There are many aspects to security, from privacy (the ability to keep secrets) and integrity through to the '3As' of Authentication (knowing who people are), Authority (allowing them to do only what they should) and Audit (the 'forensic' trace, to see what happened, who did it and when). Given this breadth of scope, the protection of enterprise networks from attack is a specialist field. Hence the network designer will typically need to seek advice from experts in the field—but there is a base level of awareness that they should have and a lot of measures that they can take themselves.

In this section we will briefly introduce some of the types of attack that the network designer must defend against, and summarise some of the techniques that the designer can work with to help ensure network security.

Understanding the threat

Before we can start to design security into our network, we must understand the level of threat to which the network is exposed, and the modes of attack that might be used. Once this is understood, we can look at what additional security can be built into the enterprise network to protect it against attack.

Many companies will require network designs to undergo a security audit against a formal standard, such as BS7799 or a risk assessment method such as CRAMM. This work would need to be carried out by a specialist consultant trained in these (fairly complex) methodologies.

Of course, the network designer should really be keeping network security in mind during the design process, so that the design will pass the audit process with little or no modification! To help with this, the designer should ask the following questions:

Who will attack my network and why?

- Probably the biggest threat comes from company employees. The attacker could be after personal gain or it could be malicious (e.g. from a sacked employee).

- Outside attack can come from the traditional hacker—who enjoys the challenge of attacking systems.

- The attack could come from a competitor or unhappy customer—they might simply want to shut the company's system down.

- Outside attack is also possible from criminals intending to commit a crime against the company. This would generally lead to financial gain by attacking applications, stealing data or programs or perhaps black-mailing a company (company reputations can be badly damaged if they have been known to be hacked).

- The attack could be indirect, for example, the company could receive a computer virus in an e-mail. This virus may have passed through many systems before reaching the company.

What components will be attacked? Any element of the customer WAN, LAN or end system is a potential candidate for attack.

- network elements—routers, LAN hubs, switches, terminal servers etc.,

- remote access (dial up) servers,

- operating systems—Unix, NT, Win95,

- web servers, and the scripts and applications (e.g. CGI and Java) that reside on them.

Of course, the information that resides on any of these could be valuable (or sorely missed if corrupted).

What type of attack can I expect? Different attackers will have different aims, so there is variety in the forms that an attack on a network can take. The main ones are:

- Direct attack—attacker aims to log on to a company application and use it as though they were a legitimate user, but for covert purposes. The attack might involve stealing or guessing passwords, using operating system or application 'backdoors' (unsuspecting users may download and run software containing a Trojan Horse which introduces such a backdoor—for example, 'Back Orifice' is used to attack PCs in this way) or subverting user authentication procedures.

- Denial of service—prevents a company network or application (e.g. a Web server) operating correctly. There are plenty of options to assist the hacker in their attack and they could go for bandwidth (using devices such as 'ping of death', SMURF), TCP/IP sockets (using TCP SYN), CPU power (using Land and Teardrop), memory, disk space etc.

- Loss of privacy—data is tapped in transit and this data could be used to damage the company's reputation or for criminal purposes. LAN sniffers and WAN datascopes are readily available and can be used for such illicit purposes.

- Data modification—data in transit could be modified (e.g. a purchase of £1000 could be made, but the attacker could modify the data to show that only £10 had been spent).

- Masquerade—the attacker simply pretends to be the legitimate host. This could be a Web site with a similar URL, designed to defame a company. (The hacker could also use more sophisticated techniques to divert traffic from the legitimate site to the masquerade site.) It could be a host simulator, tricking remote workers into revealing passwords.

- Information gathering—often the prelude to one of the above attacks. Basic TCP/IP tools like ping or traceroute could be used. Sophisticated scanning tools can be used to systematically search a host for security vulnerabilities (e.g. SATAN). Many hacking tools can be freely downloaded from the Internet!

What are the main methods of defence? These are easily remembered using the mnemonic 'CIAAA':

- Confidentiality—we can ensure that data is protected from loss of privacy through the use of encryption.

- Integrity—we protect against data modification by adding secure electronic signatures to messages.

- Authentication—we protect against direct attacks by security tests so that the user connecting to the system knows it is genuine. We can similarly protect against masquerade by testing that the host system we are attaching to is genuine.

- Availability—we can use resilient server designs which have been hardened against denial of service attacks. Simple techniques such as regular data back up are also important to allow speedy recovery in the aftermath of a successful attack.

- Audit—by keeping records of who connects to a host and what they do, we are better able to detect attacks and prevent their re-occurrence.

The designer can do a lot to minimise threats to the network, simply by avoiding the problem. For instance, if a large amount of sensitive data needs to be processed, it would be better to move the processing to the same place as the data rather than the other way round. Sometimes, though, there is genuine or unavoidable risk and this needs to be countered.

There are a number of methods that can be deployed to give some level of assurance. The remainder of this section looks at network design techniques used to implement these methods.

Network security solutions

The first question the designer must ask is at what layer in the OSI 7 layer model should security techniques be implemented? The best security will be achieved when techniques are implemented at multiple layers in the stack. This may involve multiple levels of access (e.g. one password to get onto a server, another to get into a particular application) but the odds of a hacker getting past several layers of protection are considerably less that those of guessing one password.

Before looking in detail at the nuts and bolts technology that is available to implement specific security solutions, it is worth reviewing where security can be inserted—the various layers of a network that can be protected from attack.

Application security

The most effective security will be designed in at the application layer.

- Confidentiality. Application software could encrypt data for storage and/or before transmission between client and server.

- Integrity. Application software can add electronic signatures to messages.

- Authentication. Users will be required to log on to operating systems and applications. This is often through the use of an account name and password. Security is improved where rules on password length, design and lifetime are implemented—making passwords difficult for the attacker to guess. Security is improved when the password is used in conjunction with a token, something the user knows and something the user has! The token could generate a secure sequence of pseudo random numbers, time synchronised to a process at the host site (e.g. the Security Dynamics SecureID card). It could use a challenge response principle where the host presents a random number to the user which is entered into the token. The token encrypts it and the user returns the result. The host will share the token key and can check the validity of the token.

Increasingly important in this field are Public Key Infrastructures (discussed later in the chapter). Here the token is a public key certificate (perhaps held on a smart card). This can not only be used to authenticate clients or servers, but also is important in implementing encryption and digital signatures.

- Availability. Application software has to be backed up regularly. Ideally the application is available from multiple servers at different sites.

- Audit. Operating systems and applications would be used to keep such records.

The network designer frequently does not have responsibility for selection or design of applications. Often, security in the application is not sufficient and therefore has to be enhanced by the network. This might be because legacy applications are used which do not have adequate security designed in for use in a networked environment. It may be that it is not cost-effective to provide sufficiently powerful host systems to handle the processing demands of cryptographic security.

Session layer security

Generally, secure applications will be session oriented in nature, with the user system setting up a session binding with the host system. Checks such as periodic status polling and message sequence numbering can be used to ensure session integrity (i.e. detect if the original user has been cut off in mid session and an interloper is now posing as the original user). Additional security can be applied at session start up time. This may be by means of a challenge system, where the host system sends out a random number which must be encrypted by the user system under a secret key. The response is checked to see if the user is a key holder. Other systems require the user to possess a token which generates one-time passwords, according to an algorithm which the host system can track, so it will be anticipating the next one-time password to be offered.

Increasingly, networks are using standardised security servers which will vet users wishing to access end systems and will supply them with a cryptographic token to be used when logging on to an application. Kerberos, the security mechanism within the Distributed Computing Environment (DCE) is an example of this approach. If such a security server is to be used, the network designer will have to provision network connectivity for it, and ensure that all possible users can route traffic to it. The server must have ample bandwidth to cope with peak demands (such as when workers log on to systems at the start of a working day).

Network layer security

Many network layer protocols provide useful security mechanisms. A few of the more common and useful examples are:

- Private networks and virtual private networks. The majority of security attacks are likely to come from public data networks such as the Internet or X.25 networks. Where the enterprise network has no need of inter-company communication, or is not offering a public service and very high levels of security are sought, this may be a justification for setting up a private network (based on leased point-to-point circuits and privately owned switches).

- PVCs. Networks such as Frame Relay and X.25 can offer Permanent Virtual Circuit connections. These are set up by the network supplier at subscription time. Because end users cannot set up or change the routing of these connections, it is difficult for hackers to gain connection to hosts connected in this way. There is a weak spot, though, as networks offering dial up or Switched Virtual Circuit connections give the possibility that the hacker could connect to an unprotected host.

- Closed user groups. These arrangements are found on switched networks such as X.25 and SMDS. When calls are set up, the user can quote a closed user group number. The network administrator will have agreed with the network provider just which network connections are members of the closed user group with this number. The call cannot be directed to users outside the closed user group and people not in the closed user group cannot call into it. (In fact closed user groups can take many forms to suit user needs. The network designer is best advised to discuss possible options with the network supplier.) These closed user groups rely on tables programmed into the network, which only the network supplier can change.

- Calling party identification. In networks such as X.25 and ISDN, when a call is set up, the call request will contain the network address of the calling party. The network will ensure that this address is the correct address for the physical circuit from which the call came. Host systems can maintain a list of acceptable source addresses, and discard calls from any other address.

- Firewalls. This is a form of security generally associated with enterprise networks that are interconnected to the Internet. This form of security is also used where some parts of an organisation (e.g. the finance department) need to have their applications and data protected from general network users within the organisation.

  In its simplest form, a firewall would consist of a router configured with an access list which would permit traffic only from predefined locations to pass through the router. Access lists may also be programmed to allow traffic to be onward routed to specified hosts.

  A more sophisticated form of firewall works at higher layers of the OSI stack, and normally consists of a Unix host, with firewall software, through which traffic must pass. This firewall may provide a number of functions including:

  - encryption;
  - session management with user password verification;
  - address translation gateway (where the target network does not use registered Internet addresses;
  - proxy agent—this is used with applications such as Web servers, where each access request is screened to see if the page selected is available for access by that particular user, and if so will fetch the page from the server before forwarding it back to the requester.

  A more general introduction to firewalls is included in the next section.

Data link layer security

A commonly used mechanism at the data link layer in TCP/IP based networks is the PPP protocol. This standard includes a security mechanisms called CHAP and PAP, which can be useful when using dial up connectivity. As part of the link establishment process, the calling party sends out a password which is checked by the equipment receiving the call. This exchange is done quite transparently to the user. Use of CHAP has in common with the use of Calling Party Identification a significant administrative overhead—to ensure that all possible users have their passwords or addresses programmed into all possible items of equipment that may receive the call.

Security technology

There are many devices that can play a part in protecting a network and many specialist suppliers of security advice and equipment. There are also a number of people who are willing to break into, or otherwise violate your network—a useful test of the robustness of any network security solution. In this section, we describe the fundamental concepts of security technology—encryption, Public Key Infrastructures, firewalls etc.

Encryption

The primary role of encryption is to ensure the privacy of data when in transit across public networks. One of the most effective ways to protect a network would be to encrypt all data flowing over it so that no intelligible data is available to the hacker.

Ideally, the data should be encrypted in the end system hosts and terminals. This is not always possible, and a compromise solution is to encrypt data within specialist routers or standalone encryption hardware at the end sites. Most standalone encryption systems available on the market are designed to operate over point-to-point synchronous data links—but an increasing number of systems are being developed that are able to work over packet oriented networks (X.25, Frame Relay, SMDS). The packet oriented encryptors are also known as 'payload' encryptors, as they only encrypt user data, leaving protocol headers in the clear, so that packets can be properly sent across the network.

Successful use of encryption technology will depend on secure housing of the encryptors and proper management of encryption keys (including changing them on a regular basis). There are two basic approaches to key management. The first is symmetric encryption where the security of the encryption depends on a shared secret that only the two communicating parties know. The International Data Encryption Algorithm (IDEA) and Data Encryption Standard (DES) are examples of private key systems. The alternative approach is asymmetric encryption where a user has a pair of keys—one private and one public. A message encrypted using the public key can only be decrypted using the private key and vice versa. So you can receive messages from anyone who knows your public key (which you decrypt with your private key) and can happily send an encrypted message to anyone whose public key you know. The best known of the public key systems are Diffie Hellman (used exclusively for symmetric encryption, shared secret exchange) and RSA (so called because it was devised by Rivest, Shamir and Adleman).

For all their worth, there are some major barriers to the use of encryption. The main one is actually obtaining the necessary hardware or software that removes the (sometimes) significant processing burden of encryption. Because encryptors can have military use, their sale and export is rigorously controlled. Perhaps the best known example of such restriction is the US Government's Data Encryption Standard (DES), which is classed as 'munitions' and thus is subject to export restrictions. The network designer will have to fully investigate these licensing limitations before firming up on the choice of encryption technology that is proposed in a design or in a bid.

In terms of usage, symmetric systems are currently the most popular. Because these use the same key used for encryption and decryption they operate on blocks of data and subject them to a pattern of bit permutations and logic operations controlled by the key value. As such they are able to operate at high speed and can be implemented using silicon hardware.

Within some industries (e.g. the finance sector), the preferred symmetric algorithm is the 56-bit DES algorithm. In practice, computer technology now exists to break 56-bit DES in a matter of minutes and there is increasing demand to user stronger encryption (such as triple DES) which is expected to remain safe from attack for many years to come. 56-bit DES can be supplied to any customer (other than those based in a restricted list of 'Terrorist' countries). Export restrictions currently limit triple DES supply to the so-called 'FISH' organisations (Finance, Insurance, Subsidiaries of US companies and the Health sector when protecting patient data).

The latest standard aimed at providing secure, encrypted tunnels across the public Internet is IPSec. Encryptors are loaded with sets of source and destination IP addresses or subnets, each set constituting an intranet or VPN. Unless specifically configured to do so, the encryptor will not permit traffic to pass between intranets or VPNs, and will use different sets of encryption keys for each separate network. In terms of implementation, IPSec is basically a DES IP payload encryptor, with added MD5/SHA integrity checking. Key management is effected by the Internet Key Exchange (IKE) protocol, which uses Diffie Hellman and RSA algorithms.

Less common at present are public key systems, which use asymmetric algorithms with different encryption and decryption keys. These algorithms tend to be slow, and are generally only used for signing documents and encrypting symmetric data encryption keys. Diffie-Hellman was the first such published algorithm (although it is suitable only for key exchange). RSA is the most widely used standard here today, with Elliptic Curve Cryptography (ECC) an emerging alternative. All of the algorithms are based upon complex mathematics, using problems that are known to have no algorithmic solution (e.g. RSA uses the known difficulty of finding the factors of a the product of two large prime numbers). These problems can only be tackled by 'sledge hammer' techniques that require time exponential in relation to the key length to solve. A major factor in the choice of a pulic key systems is efficiency—performance constrained systems need an efficient algorithm. In this respect, ECC appears to be the long-term favourite as it offers a higher level of security per bit of key length.

Depending on the solution used, encryption technology can also provide a number of other valuable services. IPSec encryptors tend to provide the richest feature set, covering many of the defence areas mentioned earlier:

- Access control—the encryptor can maintain lists of allowed subnets—only data to and from these will pass through the encryptor.

- Authentication—confirmation that the data comes from the location it claims to come from.

- Integrity—the data has not been corrupted or tampered in transit, through the use of encrypted message digests.

- No replay—sequence numbering and time stamping can be used to prevent improper replay of a transaction.

Public key infrastructures

Asymmetric encryption algorithms form the base of Public Key Infrastructures. As indicated above, public key schemes are widely used for electronic document signatures (to prove source and provide non repudiation of transactions) and for exchanging data encryption keys. A user will generate a public/private key pair, and publish the public key. However, we do need to know that the public key belongs to the right person and not to an impostor. The Public Key Infrastructure achieves this by having a trusted third party sign the user's public key, after the user has proved their identity to the third party. This signed key is called a certificate, and they are issued by a Certificate Authority (such as Verisign). The certificates need to be accessed from a directory, so that a message recipient can retrieve reliable public keys.

A certificate itself is basically a set of data elements, bound together and electronically signed by a trusted CA using its closely guarded private key.

A basic class 1 certificate binds a user's name with their e-mail address and their public key. This is used by individual Internet users to send secure e-mail or to identify themselves to Web servers.

A class 2 certificate is issued by an organisation such as a bank, to identify its customers. It has bound into it further details such as account numbers. The certificates are still issued by a CA, but the user's applications are processed by an in-house Registration Authority (RA), which is used to approve requests for certificates (e.g. once credit checks and written application for service is received). BT TrustWise Onsite is one of the UK providers of an RA facility.

Web server operators will seek a class 3 certificate. In this instance, the CA will carry out rigorous checks to confirm the identity of the server owner and that they are a creditable organisation. The certificate binds the servers URL, the organisation name and its public key. If a server has such a certificate, browser users can confirm they are communicating with a genuine server and not an impostor. More importantly, the certificate enables encryption keys to be exchanged and secure HTTP sessions to be run (e.g. for e-commerce and electronic banking applications).

One important design issue with PKI is where the private keys should be stored. If on a PC, the key only really proves a message came from the PC, not the person—simple user passwords are used to protect the private key on the PC, which is a weak protection. A much better option is to keep the private keys on a smart card that moves from machine to machine with the key owner. All processing operations involving the private key take place on the smart card, and there is no facility provided to extract the private key from the card at any time. When small secrets (keys) are used to protect big secrets (data), this is an altogether better design option—rather like the idea of moving the processing to the secure data, mentioned earlier.

Firewalls

One of the most effective and widely-used strategies for preserving security is to use a 'firewall' system. Just as a real firewall is a specially built barrier which stops the spread of a fire within a building so a computer firewall tries to limit the extent of damage to computer security. The basic idea is that machines and networks on the 'inside' of a firewall are trusted, those 'outside' are generally not trusted, as shown in Figure 6.9.

All communication between machines inside and outside the enterprise go through the firewall (which, in practice, is usually a special purpose computer such as Cisco's PIX or Sun's Netra). It is the job of the firewall to monitor and filter all traffic passing through it and only to allow through known communication corresponding to certain well-defined services or to trusted external systems. Hence, incoming traffic can be screened to allow access only to the privileged and outgoing traffic can be limited if there are certain places an organisation does not wish its people to visit!

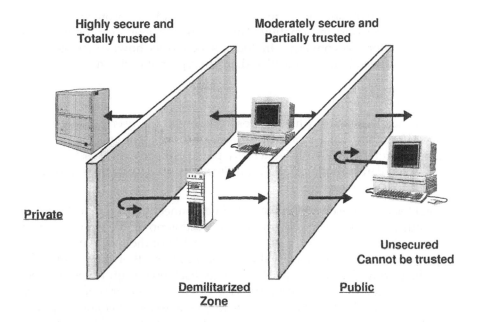

Figure 6.9   The Basic Functions of a Firewall

Machines on the inside of a firewall have a reasonable degree of trust of other machines within the firewall and so have to apply fewer controls themselves. Often, sections of the internal network will themselves employ firewalls to restrict the domain of trusted machines even further. This arrangement is usually applied within an enterprise network to give layers of protection to one organisation's information. With the growth of extranets and Community of Interest Networks (COINs), a similar approach is used to create Demilitarized Zones (DMZs), where the information shared between organisations is protected from the world at large but is still outside any one organisation's intranet or VPN. The functionality of IPSec is ideally suited for implementing secure COINs/extranets.

Firewalls are most often applied at the level of network transport (i.e. the protocols used to transfer data between computers—TCP, UDP, IP etc). However, the same idea is also applied to higher-level services with good effect. For instance, a particular service may be shielded by a proxy server. The proxy presents the same interface to the outside world as the real service but instead of performing the requested operations itself it filters out undesirable requests and only passes approved requests on to the real server.

Firewall products are increasingly able to provide encryption and this is used to create secure tunnels over public networks such as the Internet (to create IP VPNs). Encryptors are increasingly being given strong abilities to decide whether to drop, pass in clear or encrypt, based on rules. These rules would cover source address, destination address, protocol type and transport layer port information. Thus, there is a tendency for firewalls and encryptors to converge. In today's networks however, it is quite common to need both a firewall and an encryptor at each site!

Remote access security

Remote access services (that is, those that entail the user dialling into a network) are particularly prone to attack. We have mentioned some of the means of protection already—such as a password protected session log on and Calling Party Identification. A couple more useful techniques associated with dialled networks are:

- Dial back. The user will make a dialled connection to the host and go through some initial password log on. At this point, the host will clear down the call. The host will maintain a table of the user's names and passwords, associated with the phone numbers that they should be calling from. The host then calls the user back on the approved number to permit the session to continue. This can be a good form of protection where Calling Party Identification is not possible (e.g. analogue telephone networks). It is not of use where the user population is likely to be mobile (and hence the phone number they are calling from is not known).

- Tail gating prevention. A major risk area for dialled access is when a legal user completes a call. If the network is not properly designed, it may be possible for an inappropriate user to connect to that port immediately afterwards, picking up the legal user's session and continuing it. This is especially likely where users do not properly log out from their host sessions and just hang up, relying on the host to detect the disconnection and close the session (naturally, users should be discouraged from doing this!).

  To prevent this problem, it is important that communications equipment is designed to respond to loss of the connection when indicated by the modem or ISDN terminal adapter (through loss of the 'Carrier Detect' signal). When this happens, it is good practice to disable the modem/terminal adapter for a short 'guard period', of sufficient duration to permit the host to detect and fully close the application session, before permitting the next call.

  In terms of technology, many of the options explained earlier in this chapter are useful for protecting dial access. In particular, IPSec is great for securing dial up access due to ready availability of software implementations.

This brief overview has explained the security issues that the designer should be aware of and the main defences that can be deployed. The key message is that security is an increasingly important aspect of the logical design of a network and should be allocated a significant amount of attention. There are several excellent references on network security and some of the best are listed at the end of the chapter. In particular Garfinkel and Spafford (1997) and Oppliger (1998) are two of the 'must read' references for the seriously involved (and Stoll [1989] is an interesting guide if you are simply interested in the area).

6.7 SOME NOTES ON LEGACY SUPPORT

Routers can be compared to Swiss army knives, offering a variety of tools and techniques which may be helpful in supporting legacy hardware and protocols. Most legacy systems do not of course have LAN technology interfaces—rather they use serial interfaces. The first step in supporting legacy then, is to ensure that sufficient serial ports are available on the router systems. The two most common techniques for legacy support are encapsulation and protocol translation. As techniques and capabilities vary between different makes and models, the potential user is recommended to check specific manufacturer's literature, and preferably carry out lab testing before committing to large-scale deployment of any legacy support technique.

Encapsulation (also known as serial tunnelling)

This technique is used with a wide range of frame-based protocols. A TCP/IP session is established between end point routers—this session is often referred to as a tunnel. Frames are received at the end point, and placed in the user data field of a TCP packet. This is then delivered to the destination router, where the frame is extracted from the TCP packet and delivered to the end system.

Some of the protocols which may require support in this way are as follows:

- SDLC based SNA. SDLC is a polled protocol, and end systems periodically exchange frames even when there is no data to send. As an alternative to simple encapsulation, the router is often able to 'spoof' these polls locally, and by using these local acknowledgements, the amount of traffic on the wide area network can be reduced. SDLC can also be used as a multidrop protocol. The router will support this by having multiple TCP/IP sessions terminated at the 'master' end. Some routers are able to perform similar tricks for the Bisync protocol. See Figure 6.10, top diagram.

- SNA LAN based protocols. Strictly, these are not legacy protocols, but important, mainstream, modern protocols. The reason they are mentioned here is that key protocols such as LLC2 and NetBios/NetBEUI do not have a network layer protocol which a router can directly use to handle traffic of this sort. These protocols have been developed to work in enterprise networks by using Remote Source Route Bridging (RSRB). Routers are able to simulate RSRB, and in doing so, encapsulate the RSRB traffic within TCP/IP sessions. The standards based approach to this is known as Data Link Switching (DLSw). See Figure 6.10, central diagram.

- X.25. This traffic can be tunnelled on a point by point basis, or the router network may actually perform a full packet switch function, by selecting a tunnel according to the X.25 destination NUA in call requests. Some users of IBM systems have found it convenient to transport the SNA protocol via X.25. In this case, the SNA has been encapsulated into the X.25 protocol using a protocol called QLLC. The Host FEPs are generally X.25 attached, and use a software package called NPSI (generally pronounced 'Nipsy') to terminate the X.25 and QLLC. Some routers are able to support QLLC in conjunction with their X.25 capabilities.

A good use would be to attach the router to the existing X.25 circuits at the FEP site, and use the router to terminate the QLLC and deliver the traffic to the host as LLC2 token ring traffic. This is a useful early step in

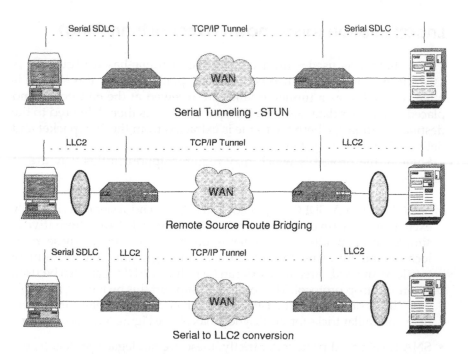

Figure 6.10 Some Options for Encapsulation

introducing a router based enterprise network into an existing SNA environment. It preserves the existing legacy investment, while allowing the host FEPs to be upgraded to Token Ring working. It removes the processing overheads and costs associated with running NPSI on the FEPs.

- Proprietary HDLC based protocols. In principle, such traffic can be carried by encapsulating each frame and forwarding it via a tunnel. Testing for each proposed application of this 'generic' encapsulation is highly recommended.

Translation

In translation, the protocol at the source end is completely terminated, and the user data is placed into a session of an entirely different protocol—one which is more easily transported over the wide area network. Inverse translation can also be carried out at the destination end, representing the traffic in its original form. Some examples of translation are:

SDLC to LLC2 translation

IBM host FEPs are increasingly being upgraded from serial attachment to Token Ring attachment. These Token Ring attached FEPs communicate with cluster controllers using the LLC2 Token Ring link layer protocol. Cluster controllers at remote sites may still have serial connection to the router. The router would be configured to terminate the SDLC session, translating it to an LLC2 session. A limitation of this approach is that LLC2 is not a routable protocol, and must be conveyed to the host FEP via a Remote Source Route Bridge network. Many routers are able to carry RSRB traffic, this generally being done by encapsulation in a TCP/IP session. See Figure 6.10, bottom diagram.

Terminal servers

Terminal servers allow asynchronous terminal traffic to be transported across a LAN interconnect based enterprise network. The terminal server performs a Packet Assembly Disassembly (PAD) function, taking individual characters and batching them up into wide area protocol packets. Data is extracted in wide area protocol packets and fed character by character to the terminal. In the LAN interconnect type of network, the most common wide area protocol is Telnet, used to interface async terminals to TCP/IP hosts. Other possibilities are LAT (used to connect async terminals to DEC hosts) and the Triple X PAD protocol associated with X.25.

DEC LAT

LAT (Local Area Transport) is a common protocol in the world of DEC VAX computers. The VAXs are frequently interconnected via Ethernets. The host applications are often designed to work with large numbers of character based terminals, which are attached to the Ethernet via a terminal server. LAT is the protocol used to carry the terminal traffic between the terminal server and the host. This protocol was designed to work over a local Ethernet segment.

Once a terminal session is established, the host and terminal server handshake many times a second. The session will be terminated if the handshakes are not responded to with in a very short period. This outcome would certainly happen if we attempted to bridge the traffic over the wide area (unless we are fortunate enough to have very high speed multi-megabyte WAN technology and the distances between sites are not too large). The best way is to completely terminate LAT sessions at each end. The traffic is typically now carried across the wide area as a Telnet session, using TCP/IP protocols.

A useful side-effect of this form of LAT translation is that as well as LAT to LAT working, it now becomes possible (by passing through the translator once only) for LAT terminals to communicate with TCP/IP hosts. Hence, terminals which normally Telnet to TCP/IP hosts can exploit LAT host services.

Problems with encapsulation and translation

The key point of concern with encapsulation or translation is the processing strain that it places on a router. Running a TCP/IP session is processor intensive (e.g. every datagram has to have an error check code calculated and checked by software—compare this with X.25 error checking, where the calculations are invariably performed in link layer HDLC serial controller hardware!). The processor-intensive nature will lead to a constraint in the maximum number of sessions that one router can handle.

The exact number of sessions will depend on the power of an individual routers processor and also the traffic levels associated with each session. A typical router might have difficulty in handling much more than 50 TCP sessions. This is unlikely to be a problem at branches, but this becomes a major design issue at the host site where many hundreds of TCP sessions will terminate. The fix is simple—if expensive. The host site should be designed so that the wide area network connections terminate at a number of routers that are reserved for IP switching only. Traffic is then directed to a number of dedicated 'encapsulation routers' whose processing power is used purely for terminating TCP sessions. The number of these routers can be scaled according to the number of sessions. This is shown in Figure 6.11.

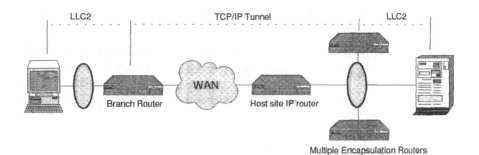

Figure 6.11 Multiple Encapsulation Routers

The other key factor of concern is the wide area network efficiency of encapsulation and translation. It must be remembered that every frame encapsulated will attract 20 octets of TCP header, 20 octets of IP header and also wide area network header (Frame Relay is one of the more efficient in this respect, adding only 6 octets of overhead). That is a total of 46 octets for every frame sent.

This problem can become more extreme in the case of character-oriented traffic. Many applications operate in what is known as host character echo mode. In this mode, every individual character typed is forwarded to the host before being echoed on the user's terminal. Such traffic is a real headache on the wide area, as every character typed now attracts the full 46 octet header in each direction. This is not only wasteful of WAN bandwidth, but the user gets a very sluggish response from the terminal.

Where possible, systems should be set up to echo characters locally, and also buffer up typed characters into full packets before sending to line. One thing that can help is TCP/IP header compression. This works on the basis that much of the data in the 40-octet TCP/IP header does not vary from one frame to the next, and hence there is much scope to compress this data. Many routers are now getting this capability.

6.8 DESIGN DOCUMENTATION CHECKLIST

We have now made considerable progress on the design of our enterprise network, and it would therefore be a good time to think about how we are going to document its design. The network designer has a clear duty to document the network in such a way that those responsible for implementing and operating the network fully appreciate the design and how it works. The following checklist shows the key areas that must be covered by the design documentation—it is also a useful checklist to help ensure that all major aspects of design have been covered.

❑ Requirements. This is the starting point for any design, and it helps to understand a network if we know what it is supposed to do!

❑ Architectural design. A high level overview of the design, showing the types of sites, their internal topology and the WAN technology and topology that will be used to link them. Note that provision of high quality diagrams is essential in the assistance of network trouble shooting and change management—there are many good tools on the market specifically aimed at network documentation (e.g. NetViz or Visio) and use of one of these tools is recommended.

❑ Physical design. This is required for each site type. This will include:

- A site diagram.
- A description of the WAN connection(s) to be provided.
- A description of the router and any other hardware to be supplied.
- A description of any racking and cabling required at the site.

❑ Logical design. This includes:

- The address scheme and naming conventions
- Schedules of assigned site addresses (usually held in a project database).
- The routing protocol design. Diagrams will be needed if it is a complex multi-area or multi-AS design.
- Resilience mechanisms—we need to understand exactly how techniques such as ISDN backup will be triggered, how this interacts with the routing protocol and how the network reverts to the primary path once it has been repaired.
- The WAN configuration (e.g. for Frame Relay we might describe how the sites are to be interconnected by PVCs, and what Committed Information Rate each PVC has).
- The software requirements—what versions and levels of code are required on each key network component.
- Protocol support—for each protocol to be transported, exactly how will this be supported.
- Describe any security aspects in the design—the threats being countered and the measures adopted.

❑ Management design

- Describe network management systems to be used by the network provider.
- Describe network facilities available to network customer.
- What network statistics and reports are required and how are these produced?

❑ Core network. What changes will be needed to the network providers core to support this network? Are there any optimum ways to connect sites to the network or other rules regarding core network use?

❑ Design verification. Here we are trying to prove that the design will meet the requirements, prior to implementation. This will include:

- – Conformance Statement: is each requirement met—Yes/No?
- – Analysis of network capacity and scalability.
- – Analysis of network delay and throughput performance.
- – Availability calculations and component failure analysis. What happens when each type of network component fails? How is it detected and how does the network repair itself—if at all? What human intervention is necessary to resolve problems?
- – Risk analysis—what unresolved issues are there with the design and what is being done to mitigate any associated risks?
- – Lab testing and trial results—describe any practical testing that has been done to verify the design.

❑ Design validation. Proof after implementation that requirements are met.

- – Integration and acceptance testing—intensive testing performed on network testbed and at initial pilot sites.
- – Commissioning tests—done at every site during roll out.

❑ Technical implementation plan. Describe the sequence of technical activities needed to implement the network—this will act as the key engineering input to the overall project plan.

❑ Futures. This is the technical strategy—how the network is expected to evolve into the future.

6.9   SUMMARY AND KEY ISSUES

This section has covered the key aspects of logical design—how we configure the physical elements of the network to make them operate as a whole system, supporting the user's end-to-end applications. The starting point for this work must be the logical design of the naming and addressing scheme for the network. It is an important task to assign a network administrator at an early stage to manage this process and subsequently to own and assign addresses from the scheme. Some of the problems associated with this can be:

- • The user may wish to use a registered IP address in anticipation of Internet connection. It is likely to be difficult to obtain other than a class C address, and this is most unlikely to meet the needs of an enterprise network. The designer will probably opt for a non registered address scheme recommended by RFC 1790, and use a gateway router to map the addresses onto a pool of registered class C addresses for Internet connection.

- Users are increasingly mobile. This has traditionally been a problem for IP networks, as host addresses are network dependent (i.e. contain information about which physical network this address belongs to). When a terminal is moved to a different LAN, the IP address programmed into that device will be inappropriate, and communications will not work. A protocol called DHCP has been introduced recently to accommodate this. A server on a LAN segment will keep a pool of IP addresses, and will assign one to a terminal connecting to the network (provided of course that the terminal can support this protocol).

- We have seen that to make modern networks function, we need to provide a number of PC based servers running Microsoft NT or Unix. These provide support services essential to the operation of the network. Examples are DHCP, DNS and WINS servers. Costs for these must be included in the network design.

The next major decision to make is which routing protocol to select, and there are many to choose from. Obviously the first consideration is that the routing protocol must be appropriate to the networking protocols that it is to establish routes for (e.g. OSPF is used for IP and IS-IS for OSI Connectionless Network Layer protocol). Choice then revolves around issues of scalability, bandwidth impact, convergence time and processing power available in routers. Different routing protocols also offer different functionality, such as ability to support load sharing. Two key issues in designing with routing protocols are that:

- We have to be aware of the scaling issues of these protocols, and these will inevitably impact the logical topology of large enterprise networks— generally resulting in a hierarchical, multiple routing area network.

- We also need to consider how each protocol on the network is to be supported. They can be carried natively, or encapsulated in another protocol (usually TCP/IP). The key here is that in a large network it is often best to restrict the core to a single protocol that routers can switch most efficiently (i.e. IP). Other protocols can then be carried over this IP core by encapsulation.

For the great majority of commercial network users, security is always high on the agenda. We have described a number of techniques that can be used to help this. Key issues are:

- It can be hard to balance the need to protect these networks from intruders, without making life unnecessarily difficult for legitimate users.

- For the highest levels of security, encryption will be the method of choice. However, government restriction on the export and use of encryption equipment has to be considered, and it may not always be possible to use this technique legally.

Finally, we discussed support of legacy equipment. There is often considerable corporate investment in these systems and their protocols. Fortunately, routers can often provide an answer, through the use of encapsulation or translation. The key issue here is that encapsulation and translation are processor intensive techniques, and there are always design constraints on the number of connections or sessions that can be supported by a router. Designs may require additional router hardware dedicated to this function.

This last point illustrates the fact that logical design can (quite often) have feedback into the physical design of the network. In practice, the designer needs to have a good top level understanding of how the logical design will work before the physical design work can be completed.

REFERENCES

Comer, D. (1995) *Internetworking with TCP/IP Volume 1 Principles, protocols and architecture* (3rd edn). Prentice Hall.

Frost, A. and Norris, M. (1997) *Exploiting the Internet—Understanding and Exploiting an Investment in the Internet.* John Wiley & Sons.

Garfinkel, S. and Spafford, G. (1997) *Web Security and Commerce.* O'Reilly.

Norris, M. and Winton, N. (1997) *Energize the Network; Distributed Computing Explained.* Addison-Wesley.

Oppliger, R. (1998) *Internet and Intranet Security.* Artech House.

Stevens, W. R. (1994) *TCP/IP Illustrated Volume 1—The Protocols.* Addison-Wesley.

Stoll, C. (1989) *The Cuckoo's Egg.* Doubleday.

BIBLIOGRAPHY

Cheswick, W. and Bellovin, S. (1994) *Firewalls and Internet Security: Repelling the Wily Hacker.*   Addison-Wesley.

OSPF Network Design (1998) Cisco Press.

7

Operating the Network—Network Management Design

We should distrust any enterprise that requires new clothes

Henry Thoreau

It is one thing to build a network, another to get the most from the time and money that you have invested in it. To paraphrase Churchill, the design of the network is not the end or even the beginning of the end. It is the merely the end of the beginning.

If the enterprise network is to provide a reliable service to the users year after year, effective network management is essential. The design of such systems can be complex, and deserves as much attention as the design of the end-to-end application data path.

The real effectiveness of an enterprise network depends on how well it performs—day after day, month after month, year after year. An expensive resource that is central to a company's operations needs to be taken good care of; that means both maintaining the equipment and ensuring its continued relevance. The former is usually called network management, the latter is the essence of service management. This chapter covers the basic techniques and tools that underlie good practice in both of these areas.

A first point to make is that the real problem that is being addressed in service and network management is the control of complexity. Modern networks are simply too diverse and disparate to be controlled by anything other than well thought out principles and supported by powerful tools;

and this is not something tackled as an afterthought. Management capability has to be built in as an integral part of the network design.

In large enterprise networks, especially those where the operation of the network is outsourced, the inherently complex management picture is further complicated due to the need for multiple management systems. These are:

- Network Operator Management System. This is a fully capable platform able to provide operational support of all the outsourced network elements.

- Network User Management System. This allows the end user an overview of the status of the outsourced network—usually by delivery of a 'read only' view of information obtained form the network operator management system. This view is required by those in the enterprise with responsibility for operating the end-to-end applications and the associated end systems. This information can then be integrated with information from the users end systems to help localise reported faults to end systems or the network.

The key focus of this chapter will be on the systems that must be designed and installed to enable day to day operation of the network. These systems must achieve the following:

- Operational Assurance—provide evidence of correct operation of the network, achieved by the periodic polling of network components to establish they are operational and reachable over the network.

- Alarm Handling—processing of alarms (also known as traps) from network elements, highlighting network faults.

- Diagnostic Support—tools to allow the user to investigate and resolve faults.

- Statistics Gathering—to provide Quality of Service reports to managers, assist in network planning and help in proactive diagnosis of developing fault situations.

- Configuration Control and Image Server—this is the central development, storage and downloading of network component configuration data. The image server is required to download router software, when this must be upgraded, usually using the TCP/IP Trivial File Transfer Protocol.

There are many other aspects of network and service management that must be considered. Space permits only a passing mention of some of them here:

- Delivery—this is the effective handling of requests for service. Effective database and project management tools will be required to support these.

- Inventory Control—this means keeping a track of the physical components delivered into the network, address allocation and the topology of component interconnection.

- Network Audit—tools to check that network actually matches information held by the inventory and configuration control systems.

- Help Desk—the end user must have a well-defined point of contact for handling faults and queries. The help desk team will need ready access to the management systems described in this chapter.

- Fault Logging and management—a database system is needed to log reported faults. The system should automatically prompt users for status updates and escalate faults not fixed within target times. The system should also provide statistics on fault causes and clearance times.

There are a number of well-established models of communications management that provide more detail to the above list. Perhaps the best known being ISO FCAPS (Fault, Configuration, Accounting, Performance, Security) and The Telemanagement Forum's Telecomms Operations Map which defines four layers of communications management (Business Management, Service Management, Network Management and Element Management).

Although useful, we will not explore these models further but describe some practical management systems that illustrate the main issues tackled in them. First, though, a few of the problems that we are seeking to cure.

7.1 WHAT GOES WRONG?

It would be nice if a well-designed network, once installed, just kept on working. But this is never the case, even in stable operation there are still plenty of needs a network management system must meet. The potential problems that network management design should cater for include:

- Chain reaction failures. Where you need to know how a failure in one part of the network might affect the total operation. For instance there may be a bug in the database software that keeps track of network addresses. It could block access to critical services like back-up systems, but at the same time it could hide this problem from the monitoring system. You need cross-checks and consistency checks to detect this kind of problem.

- Traffic congestion. Just like a well-designed highway, any network can suffer from traffic jams. If several network elements fail simultaneously, the load of blocked and diverted traffic can bring the system to a halt. Adding to the congestion are the messages the network generates to report the problems.

- The unexpected. A network hit by unexpected events must be able to help itself. It should manage and reroute traffic to avoid trouble spots. The system must also react properly to duplicate messages or verify messages from questionable sources.

- Centralised and decentralised management. Central management can also create a central point of failure. Decentralised management can be a source of inconsistency. In an enterprise network, you are likely to have elements of both, with their combined drawbacks. You must decide who should be responsible for managing such things as database consistency, standby systems, and database updates. Another decision is who should receive status information and error messages?

- Protocol standards. The choice of a network management standard—or none at all—can either improve management or make it harder. If you base the system on standards, you must be sure the entire system follows them. Otherwise, it might interpret non standard messages in strange ways. You must also make sure you can support the standards you want to maintain. On the other hand, standards make it easier to integrate network management with the network as a whole.

- Growth potential. The management system should adapt to traffic growth and the addition of new nodes and networks. It should also incorporate new technology and accept new features as opportunities arise. This should be done relatively easily and with little disruption.

We now move on to look at the architectures of typical management platforms used to tackle these problems.

7.2 A BASIC MANAGEMENT PLATFORM

Figure 7.1 below show a minimal network management system. This consists of a Unix workstation, connected via a router and core network connection to the target enterprise network sites which are to be managed.

The element manager will typically be a Unix platform running Simple Network Management Protocol (SNMP) management software (e.g. Sun Net Manager or HP Openview). SNMP itself is a straightforward request/response protocol that allows retrieval of information from agents located in each network component. The agent accesses the data from a Management Information Base (MIB) located within the component. 'Get' commands can be used to retrieve MIB parameters, thereby discovering the status of network components, and 'Put' commands can set MIB parameters, which allows configuration of the component. SNMP also permits network components to deliver traps (also known as alarms) and have them delivered into the management system.

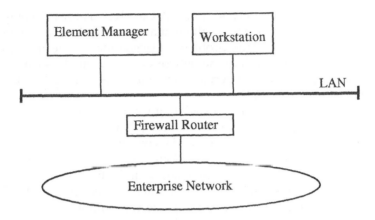

Figure 7.1   Basic Management System

The task of the SNMP management software is to provide a network manager with a graphical view of the network, highlighting nodes with problems. The user will be able to use the graphical user interface to home in on such nodes to perform further diagnosis (often through using TCP/IP Telnet to log on to the router, which will offer a much richer set of management and diagnostic facilities than can SNMP alone). The management system is also able to gather statistics and prepare reports to assist in long-term health checking and capacity planning for the network.

The element manager is connected into the enterprise network via a LAN and a router. The router will perform a security firewall function, to protect the management system from misuse. Multiple users can be accommodated by adding extra workstations, which can have sessions with the network management workstation using the X Windows protocol.

How does the management system know the state of nodes? There are two key methods:

- Alarm or trap processing. Network components tend to produce a rich set of event notification messages (examples would be failure of a Token Ring attached to a router, or if a router is running short on memory). These messages are sent out, using the SNMP protocol, and logged by the management system. The trick with dealing with such messages is to be able to tell the wood from the trees!

  The management systems will require alarm filter capability so that only the most serious events are brought to the attention of the operator. Where possible, information type alarms should be suppressed at their point of origin, to avoid unnecessary traffic—most network components will have the ability to configure just how talkative they are about their status.

- Alarms are not in themselves sufficient. Clearly, if a network component or its communications path to the management system have failed, no alarms will be seen. This problem is overcome by the management system periodically polling the network component. This can be done using a ping (a process whereby a test packet is reflected by the node under test using the TCP).

  Alternatively, we can use SNMP to get a one or more variables from the nodes Management Information Base (MIB). This is the repository of information on network devices, and it contains a description of SNMP-compliant objects on the network along with the kind of management information they provide. These objects can be hardware, software, or logical associations, such as a connection or virtual circuit. An object's attributes might include such things as the number of packets sent, routing table entries, and protocol-specific variables for IP routing. Clearly there is a design trade-off to be made here—if you poll too often, excessive network bandwidth is consumed. Polling too infrequently means that it takes too long to react to faults. Polling network components every 5 minutes is often a sensible compromise.

Network statistics are also gathered by polling. In this case, we need to make a whole series of SNMP 'get' commands to fetch the relevant data. A typical MIB will contain many kilobytes of 'useful' data. It is all too easy to consume excessive bandwidth and network management system disk space, by collecting too much too often. Collecting statistics every 15 minutes usually provides adequate resolution. Care should to taken to request only the minimum necessary information. This might be:

- Router peak processor utilisation—to help predict when and where more powerful routers might be needed.

- Router free memory—to identify where more memory might be needed.

- Network access link utilisation—to identify sites needing more capacity, or where we need to investigate how the load can be reduced through tuning.

- Packet drop rate—routers are generally dealing with connectionless IP type protocols, and they cope with port congestion by simply discarding or dropping any overload. The drop rate is a good warning that undesirable levels of congestion are being reached.

- Errored packet rate—network access links will generally be running protocols based on HDLC framing. These frames include a frame check sequence, which routers use to detect and discard faulty packets. High rates of errored frames will indicate links that need maintenance attention. Note that high rates of line errors may also raise alarms.

7.3   A MORE COMPLEX MANAGEMENT PLATFORM

While the above platform may prove adequate for managing small router networks, it is unlikely to satisfy the larger enterprise network user. Issues include:

* Scalability. One workstation, running element manager software, can typically only poll a few hundred network elements. A big enterprise network can consist of thousands of elements. An outsourcing supplier with many customers could be faced with handling tens of thousands.

* Proprietary components. Not all network components use standards-based management protocols. They will have proprietary management platforms, providing equivalent information (e.g. an enterprise network may rely on managed LAN hubs that are managed using Novell protocols and a wide area network supplier may provide a proprietary information feed to notify the user of the current state of the network). What the network manager needs is an integrated view of the whole network—not a console with a dozen different screens, each showing a different component.

* Customer view. In the outsourced scenario, the user may need a 'read only' view of the network (e.g. to assist the user's application help desk to see if a reported fault is due).

* State-of-the-art. When developing new networking technologies, it is common for the development of the management facilities to lag behind the development of the user data path. This occurs for easily understood economic reasons. The company developing a new data communications product will want to see an early return on its investment and will thus want to focus development effort toward early product launch. It is possible to sell a new technology that can transport user data in some novel and cost-effective manner, even if the product has rudimentary management ability. You could not sell a product that has excellent management abilities but does not provide effective user data transport! As enterprise networks often call for the use of state-of-the-art networking components, network management capabilities may well fall short of expectation.

The result of this is that the network designer is unlikely to find a single product that will meet all management needs, and an effective platform must be created by blending the best features from a number of products.

A typical large scale management system is shown in Figure 7.2. This consists of a number of element managers, some for standards based and some for proprietary systems. Each element manager will typically support several hundred elements, but the exact number of element managers will depend on:

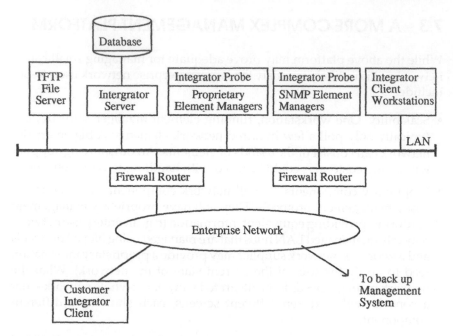

Figure 7.2 A Typical Large-scale Management System

- the capability of element manager software;
- the power of the hardware platform deployed;
- the frequency of element polling and the volume of data to be collected.

The element managers communicate with the elements they manage via the enterprise network core and firewall routers. It will be noted that in this more complex platform, redundant network connections have been used to improve the availability of connectivity at this site.

The heart of the platform is the 'integrator' software. This can be considered to be a 'manager of managers'. An example of such software is the Micromuse Omnibus product. Each element manager will contain an integrator probe, that makes it possible to 'robotically' operate the element manager, gather information about the elements and then write the data to a database server. The integrator server generates the required management views from the database for the integrator clients.

It is now possible to provide the customer with a view of their part of the network by providing them with an integrator client, attached to the server via a network connection and a firewall security router. The customer will be set up with an account on the integrator server that allows them a view of only their part of the network.

Also shown in Figure 7.2 is a TFTP file server, which is used to store network element software images and configurations. Network elements such as routers will be configured with the address of this device and can request downloads from this as necessary.

7.4   DISTRIBUTED MANAGEMENT

In the above sections we have described highly centralised management systems which rely on collection and analysis of large volumes of data for correct operation. The levels of traffic generated by this process can easily impact the performance of a network, if too much data is collected too often. Emerging techniques in distributed management systems can help considerably with this problem.

This technology works by placing management probes at the remote sites (generally those with problems or considered likely to be the busiest). These probes can either be software running on routers or LAN hubs or dedicated network analysis hardware. The idea of these units is that they capture large quantities of data about network conditions locally, and only send summary reports to the central site on request. At present, the default option would be to adopt the Remote Monitoring (RMON) standard.

These probes can also be used to generate smart alarms, through a process known as benchmarking. The reports produced by the probes are analysed by the network operator to establish what normal peak load conditions are like at that site—in terms of LAN and WAN utilisation, packet drop rate and router processor utilisation to establish operational norms. The probes are then configured with these norms as thresholds, and only raise network alarms when sites exceed thresholds. These alarms can also condition the probe to start much more intensive information gathering, so the network manager has a good chance of discovering what is actually happening at these times.

As network management technology evolves, we can expect to see increasing levels of sophistication in what can be done to monitor the health of a network at local level—the centre only contacted when things are starting to go wrong.

7.5   ISSUES IN NETWORK MANAGEMENT

Having selected and designed the management platforms, we now need to consider how and where to connect the management system into the enterprise network. There are two principal ways of doing this, as illustrated in Figure 7.3.

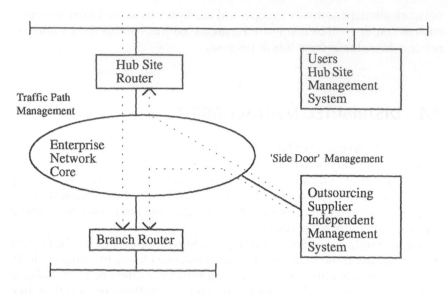

Figure 7.3 Two Alternative Management Approaches

Hub site attachment

In many enterprise networks there is often a small number of major hub sites, typically those where host computers or large corporate servers are located. All smaller sites will need network connectivity to the hubs. It is sometimes appropriate to establish a management system at one of these hub sites. Often it will be advisable to establish a second standby management system at an alternative site to allow for failure or loss of the primary site.

In a peer to peer network where all sites are similar, with any to any type communication, one site will have to be nominated as the management site—usually the location where the network operations staff are located.

In this approach, the management platform would be attached to a hub site backbone LAN from which paths can be established via the wide area network to the branch sites. The management platform will often need to be connected to the hub backbone via a bridge (or router), as there can be a very great deal of traffic between the different components of the management platform. It is undesirable that this traffic should pass onto the hub backbone, where it might cause congestion to customer application traffic.

This method of management is the norm for companies operating a private enterprise network. Also if the company has an outsourced solution based on point-to-point circuits or a TDM network, this will also be the solution adopted.

Independent site

An alternative approach is to establish an independent management site. This will have management paths built through the wide area network to each site to be managed. (This is generally only practical where the wide area network is of the statistical multiplexing/packet switching type, e.g. Frame Relay, SMDS, ATM, where the additional management path can be provided at marginal cost.)

This method of management is generally employed by outsourcing suppliers who manage enterprise networks for multiple customers. In order to achieve economies of scale, they do so from a central management site (in fact, such an operator would normally operate two such sites for the sake of resilience—each customer site will then require a management path to each of the two management sites).

Issues in selecting connection technique

Of the two techniques, hub site attachment offers a superior view of the network in terms of the customer applications. This is because management traffic will flow over the same end-to-end path as application data, giving a better indication that applications are likely to succeed than the independent site method. A disadvantage of the hub site technique is that all management traffic is concentrated through the hub site routers and WAN connections, which could impact application performance. The independent path technique means that the hub sites only handle management traffic relating to equipment at that site.

Addressing can be a problem for independent management systems. These generally require their own addressing scheme, which is normally a registered one. The reason for this is that independent management sites normally handle multiple customers, who could all potentially use the same RFC 1597 unregistered address scheme. None of the customers should be using the management registered address. This makes router configuration more complex as WAN ports will now need multiple addresses.

The port will need to be assigned a primary address from the customer's address range. This can be used by the dynamic routing protocol, thus enabling customer traffic to be routed via that port. The management address is assigned to the port as a secondary (or alias) address. This is used exclusively for management, and management traffic is normally directed to and from this address by static routing. Any traffic received at a router with a destination address matching one of the addresses assigned to a router port is taken as destined for that router. It will be passed to the routers internal TCP/IP stack and processed appropriately (e.g. management traffic will be passed to the SNMP handler task in the router).

Resilience and the issue of cascaded routers

Hub site systems are potentially less robust than independent systems. This is due in part to the fact that it is not always financially possible to build two complex management platforms. This is much easier when the platform is shared. The other reason is due to cascaded router management paths.

A router is cascaded when it is necessary for management traffic to pass through one or more customer routers to reach the target device (see Figure 7.4). Failure of any one of the intermediate routers or nodes will result in loss of contact with a cascaded router. In these cases it can be difficult to localise the problem—is it the remote router, its network connection or a problem with an intermediate connection or router?

Independent connection management systems rely much less on cascading and designers working on such networks will often aim to avoid cascading, perhaps permitting one level only, where unavoidable. This means that when two routers fail to communicate, it is often possible to still reach one or both via the management path, and use the often sophisticated diagnostic capabilities of the router to localise the problem.

The best of both worlds?

We have now suggested two possible network topologies for the connection of the management system to the network, each approach having its own pros and cons. What about a 'best of both worlds' situation, where the network is managed from both a customer hub site and also from the network outsourcing providers management centre? Such arrangements are indeed possible and commonly encountered in practice. Possible considerations in the design of such arrangement will include:

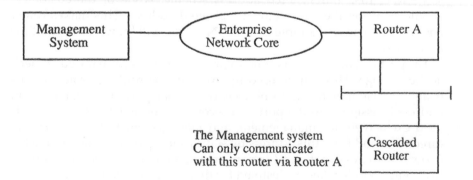

Figure 7.4 An Example of a Cascaded Router

- Role Agreement. The two management centres will have to agree who is responsible for doing what, to ensure all functions are covered, while keeping duplication of effort to a minimum. An important area here is who has responsibility and authority to make changes to network configuration?

- Management traffic minimisation. If both sites carry out extensive polling and statistics gathering, the management traffic coming and going from branch sites can account for a significant proportion of the access circuit bandwidth. Again, agreements are required as to which system should gather which information. It is often possible to set up the systems so that one system gathers the data and this then makes the gathered information available to the other site on request.

- Cost justification. This is clearly an expensive approach, as network management systems are not cheap! The cost can often only be justified where the network is carrying mission critical data. It may well be most cost-effective to leave the management to the network provider, and just have a system which provides a read only view of the current state of the network, which is supplied by the network provider.

7.6   NETWORK MANAGEMENT SECURITY

The network management system is a key area of network security risk. A malicious user of these systems, or someone who has 'hacked in' from outside can easily do significant damage to the network (say by shutting down systems) or pave the way for attacks of user end systems by disabling security filters. Techniques that can help protect the network are:

- Keeping the network management systems and their user consoles in secure accommodation, accessible only to legitimate network operators.

- Careful selection and training of management staff.

- Use of passwords for accessing these systems and for logging on to routers and other network equipment. These passwords should be changed regularly, and especially when staff leave the organisation!

- Ideally routers should be protected by security systems more elaborate than simple password log on. For example, Cisco offer a network security server approach called TACACS. When a user tries to log on to a router, he or she will be asked for a user name and an associated password. This is referred to the TACACS server for validation before log on is permitted. Since legitimate user names and passwords are held centrally, it is much easier to ensure that user lists are kept up-to-date and that passwords change regularly.

Another potential risk is introduced in the case of a networking service provider, where there is a management system shared across many customers. Great care must be taken in setting up routing in this situation. The main risks are:

- The customer's routing process may discover a traffic path via the management system. If this happens, the resulting traffic load on the management system connections could prevent the management system operating effectively.

- There is a possibility that one customer's network could discover another's. This would result in a severe loss of confidence in the security of the network, if one of the customers were to discover this.

One possible resolution of this problem is to insist that *only* static routing is used between a management system and the managed components. These management paths should be set up using a registered addressing scheme, which should never be included as a network in any routing process. If this is done, the customer's routing processes will not be able to discover a route via the management system.

7.7 SUMMARY AND KEY ISSUES

In this section, we have emphasised the importance of network management systems to the success of an enterprise network. The management system will merit as much care and attention in its design as the network data path.

Key issues to be considered in the design of these networks are:

- Scalability of the design. There are finite limits to the number of elements that a single network manager can handle—large networks will need complex management systems.

- SNMP is not the only management standard! Many useful and important networking components will have proprietary management feeds. A management platform that can integrate various feeds into a single big picture is better than multiple consoles.

- Performance impact. With centrally managed systems it is all too easy to gather too much data too often, impacting the overall performance of the network. A distributed approach to management can address this issue.

- Topology. Choosing the best way to connect the management system to the network can critically affect the quality of the information gathered and the effectiveness of the management system. In the larger enterprise network, there can be a case for multiple views.

- An application view. Even if we have designed an optimum network management system, this may not be enough. It is all too common for a network to appear perfectly healthy to the management system, but for the customer's end-to-end application to fail. Effective management systems need to look increasingly at the network from the point of view of the end-to-end application. This could be done by systems able to intelligently monitor application traffic, or by combining feeds from the host/server management systems with those of the network. In an ideal world, as soon as the management system flags that a customer site has lost its application session with the host/server, the network manager should be investigating proactively to see if it is a network problem!

BIBLIOGRAPHY

BCS Medallists report (1994) The global office. *Computer Bulletin* February.

Held, G. (1992) *Network Management—Techniques, Tools and Systems.* John Wiley & Sons.

NCC (1982) *NCC Handbook of Data Communications.* NCC Publications.

Valovic, T. (1992) Network management: a progress report. *Telecommunications*, August, 23–32.

8

Planning for the Future

Quality is optional—you don't have to survive

Philip Crosby

Successful enterprise networks will be those with sufficient flexibility and agility to evolve and meet the ever-changing requirements of its users. If, by now, we have managed to create a useful network, the traffic on it is likely to grow, and eventually hit capacity. As well as traffic growth, the network designer will be faced with many requests for changes as applications and end-system technologies evolve.

One of the essential measures of the quality of a network is its ability to change and meet new requirements. These may be driven by improvements in technology, by demands for richer functionality or by changes in the network's user base as the organisation it serves grows through expansion or acquisition.

Each of these drivers for change are likely to impact on the network and each must be catered for during the design process. This chapter aims to give guidance to the network designer on preparing for and coping with the changes required to keep up with the times.

8.1   CHANGE MANAGEMENT

An obvious (though often overlooked) requirement is to maintain control over change in the network. It is all too easy for the office maverick to add an extra server here and an extra LAN bridge there, which may result in unforeseen consequences to the network as a whole. Modern enterprise networks can be large and complex. Their stable operation depends on taking great care to ensure that many interacting design rules are not exceeded (some of these rules have been introduced in Chapters 5 and 6).

Our experience has shown that some of the (seemingly innocent) changes, such as integrating a new group of office routers into the network may have wide-ranging consequences. These extra routers may cause OSPF routing area limits to be exceeded, or greatly increase the amount of routing information that other routers in the network have to hold and process. As a result, the whole network can be made unstable and may fail entirely. Both authors have been involved in more than one 'drains up' exercise on an 'extended' network or an 'improved' piece of software, where the failure was tracked to an uncontrolled change of this type. It is the sort of messy activity that everyone should strive to avoid.

Even the network suppliers themselves can be cajoled into making such rapid and frequent changes that the design documentation is out of date, compromising the ability to manage the network effectively! A formal network change control process should be set up as an integral part of the company's quality management scheme (see section 8.7 which outlines such a process).

Changes to a network can come about not only as a result of a customer request, but also as a request from the network supplier—perhaps to fix a network fault. There will always be conflict between the need to make changes to a network and the need to maintain operational stability, so that the end users can reliably exploit the networked applications that the network is intended to support! One way to handle this is to treat the network in much the same way as a software project—in which it is not desirable to issue a new release of software for every change. It is simply too expensive and disruptive to make such frequent changes, as each change will need regression testing, followed by careful phased roll out.

The answer to this is to exercise Version Control on the network design. A version plan is produced, showing which changes will be addressed in which element of the network design. Advantages to the approach are:

- Disruption and expense is minimised.

- We can accurately describe the current state of a network (or part of a network during transitional periods) in terms of a design version number.

- Documentation can be changed and reviewed to reflect each version, and marked with a version number to help show that we are using up-to-date documentation.

- It is easier to plan a clear technical strategy that can be agreed with the customer by defining design versions.

- It is easier to agree prioritisation and timing of changes if they are slotted into a defined version plan.

It is suggested that, as a minimum, a two-part version number be used:

`<Major Version>` · `<Minor Version>`

Major version covers significant network wide changes. These would include:

- Lifecycle phases—Trial network, pilot network, rolled out network.

- Major scaling—Large numbers of new sites, capacity increase in core network connections etc.

- New site types added.

- Changes of core network technology, e.g. migration from Frame Relay to ATM network.

- Network-wide router software version upgrade to introduce new functionality.

Minor versions cover:

- Additions of small numbers of sites to the network, where new sites are similar to existing sites in design.

- Router software upgrade for bug fixes.

- Addition of extra end systems at some sites.

In creating the version plan and implementing the required changes for a new version, the network designer has to make finely balanced judgements between:

- Ensuring changes are delivered quickly enough to meet user needs.

- Maintaining network stability—not changing things too quickly and ensuring that key network design parameters are not exceeded as a result of the change.

- Not changing too much at once (if the change fails, it will be harder to identify what went wrong and to regress to the previous version.

8.2   DESIGN REVIEW AND CAPACITY PLANNING

Having examined how we can try to maintain some control over an evolving network, we will now look at ways in which the need for these changes can be proactively anticipated. We will also examine different aspects of how this might be achieved.

Clearly, change can best be handled if we know that it is coming, and can plan for it. Nothing is worse for a network operator than to receive a user change request demanding that some new facility or addition capacity is provided in an unrealistically short timescale!

An effective way to anticipate change is to hold periodic design reviews with network user representatives and application developers (a frequency of quarterly or every six months is suggested). These reviews have two key aims:

- to establish that the network is meeting existing requirements;

- to establish any future requirements.

When looking at existing requirements, we will need to determine if all quantitative requirements are being met (network delay, throughput, peak traffic loading and availability). Our management systems should be providing the evidence needed to investigate these areas. It may well be that, while these requirements are being fully met, the user community is still not satisfied. Perhaps a network delay of two seconds has been agreed but the server is adding a further two-second delay. Whatever the reason, the overall delay may have proven to be very annoying to end users, who now seek better performance.

These reviews should be seen as an ideal opportunity to gain some intelligence about upcoming requirements. Some of the main areas to probe would be:

- What trends are coming out of our network traffic figures? Suppose we know that our peak traffic time is on Friday lunchtime, we could plot this peak on a weekly basis and extrapolate likely traffic levels in the future.

- Is any growth expected in traffic from existing applications? This may be due to increased business volumes, or simply due to more people with network connections being trained to use these applications.

- Are additional sites or users to be connected? Has the enterprise expanded or acquired a new unit? If so (or even, if likely), where would the new users be, how many of them are there and what sort of network facilities do they have in place?

- What new applications are being developed? Application developers will often write new applications using a very high-level interface to underlying network communications (an example might be a SQL database query—the application developer need hardly be aware of the network at all!).

  As soon as new applications are identified, it is advisable for a networking expert to investigate it and assess the levels of networking traffic that the application may produce. It may even be possible to get the developer to change the application in such a way as to use the network with better efficiency, if a network analyst is involved early enough.

*   Identify any strategic directions that may influence the network. An example of this is the emerging trend to centralise server systems. There has been a trend for servers (e.g. file servers and mail post offices) to be sited in smaller branch offices. These systems need routine administrative support (e.g. back up and adding new user accounts). This has been done by non-specialist staff at the branch, often with high levels of support from central help desk facilities.

    It is being increasingly realised that there are high 'shadow' costs involved in this (i.e. workers are being deflected from their main role to do this support). The reaction to these costs is to move servers back to more centralised sites, where fully trained, dedicated staff can support them. This centralisation of servers is tending to require higher network bandwidth to support the traffic to the remote server.

Clearly, changes required to the network identified by this process will have to be reconciled with the network business case. Upgrades to the network will invariably involve additional expense, which must be shown to be cost justified. If insufficient funds are available to carry out all the identified changes, then work will be required to prioritise the changes, deferring some to a later date. Alternatively, application and network designers may be able to identify more cost-effective methods to achieve the same goals.

8.3   ASKING THE 'WHAT IF?' QUESTION

So far, we have looked at trying to anticipate the future shape of a network by extrapolating from past performance, and by quizzing the user community about their future requirements. It would be wise, however, for the network designer to 'always expect the unexpected'. A little bit of crystal ball gazing is highly recommended! An adjunct to this would be that you should assume the worst in any prediction—the 'law' that states that any network will exceed its planned capacity generally holds true.

When reviewing the network design, it is always wise to ask questions about scalability. A factor of 2 is definitely worth considering and possibly also a factor of 5. The following types of growth should be considered:

*   number of sites
*   volume of transaction type traffic
*   size of files to be transferred
*   number of sites transferring files simultaneously

Where the user community is not clear about future application direction, it may well be worth the network supplier engaging in some speculative thinking. For instance, there has been much hype concerning multimedia, and increasingly user workstations have the appropriate peripherals to support multimedia applications. But what are the applications? Multimedia is a technology, and if it is to be exploited, the business will have to find key business requirements that can be satisfied in a cost-effective way using this technology. Let us look at a couple of possible applications for multimedia.

Document image transfer

The multimedia workstation is now able to display high-quality colour images. It also has the processing power to handle the compression techniques that make it possible to store high-quality images in reasonable size files (e.g. TIFF or JPEG format). It is possible to store these images on a database and access them remotely. This technique will be of great interest to many business sectors:

- Banks may wish to save scanned images of cheques centrally and retrieve these at branches if the customer has a query.

- Estate agents may wish to use a central database of photographs and particulars of houses for sale.

- Shops may wish to develop electronic catalogues of goods too large to stock in a high street branch, but which customers may want to order there.

Those who have used the Internet World Wide Web will already be familiar with the ability to retrieve such images from databases around the world.

The problem with this technology in business applications is that the users will expect the same swift transaction response times that they are used to with traditional transaction-based systems, typically around 1 to 2 seconds. A compressed image is typically 50 kbytes in size. To achieve customer expectations, we would need a site connection of at least a half Megabit per second, e.g. a low speed SMDS access.

Interpersonal communication

Customers are increasingly requiring their enterprise networks to support a broad range of voice and video facilities, to replace traditional telephony and to provide video conferencing, remote training or improved employee communications channels. At present, many customers will have separated this kind of traffic onto a dedicated network (e.g. PSTN, ISDN or private circuit networks interconnecting PABXs). Technology is fast evolving that allows this kind of traffic to be transported over the enterprise data network.

The argument is that digitised voice or video is just another kind of data, and that much cost can be used by exploiting 'spare' capacity on the data network to convey this traffic. It also facilitates use of technology such as 'soft PABXs' (inexpensive PC based platforms able to replace traditional expensive PABXs—providing the voice is presented as digital traffic streams via a LAN) and computer telephony integration (where a tight coupling between voice processing and customer databases is exploited to allow fast, efficient transactions to take place in call centres).

A key emerging technology is voice over IP (VoIP). This works by sending the digitally encoded voice samples across the network as IP packets. A key issue with voice technology is that it is very sensitive to network delay and variations in network delay. The delay variations can be smoothed out using buffering techniques.

To make VoIP work, you either need a high bandwidth network with plenty of free capacity (to ensure network delays are kept very low) or a network that has arrangements for implementing some form of Quality of Service (QoS). This will ensure that the voice samples are sent over the network at higher priority than data traffic. Results from this technology can be variable, and would-be users are advised to test carefully that voice quality will be adequate, while mission-critical transaction and file transfer performance is not adversely impacted.

As ATM networks become increasingly common, it is possible to use bandwidth from this to support voice and video. ATM supports the ability to establish separate network channels for voice and video, offering this fixed bandwidth/delay properties, while data can flow on channels offering variable bandwidth/delay properties. An ATM service multiplexor is used to merge separate voice and data streams onto a single WAN bearer circuit. A common use of this technology today is to use ATM service to link the PABXs at core company sites using 2 Mbps channels taken from the available ATM bandwidth at these sites.

More details about how a typical customer might want to exploit voice over data technology can be found in the worked example in Chapter 10.

8.4 FUTURE TECHNICAL STRATEGY

User applications and networking technology continue to develop at a rapid pace. It is all to easy for today's new enterprise network to become an obsolete, legacy network within as little as three years. Enterprise network business cases should be constructed to write off much of the network infrastructure in this kind of timescale, so that the need for the funding required to migrate the network to new and faster technologies is anticipated.

It is of course for the network designer to come up with the technology strategy for the enterprise network. To assist with this, the designer can use the following techniques:

- Close component supplier relationships. This will allow the designer to get advance information on new products and the features available in new versions of component hardware and software. These should be assessed to see what benefit they might have to the network.

- Continuous education. The network designer *must* find the time to keep up-to-date with emerging technologies and products. This can be done by reading books, technical magazines and attending seminars and training courses. Seminars, organised by the professional institutions or private companies can be an especially useful source of information, since the network designer can meet with peers and find out what is happening in other organisations.

- Lab work. Where funding permits, every opportunity should be taken to trial candidate technologies in a lab environment. This will help with the process of continuing education for the designer. It will also show the *actual* state of candidate technology, much of which may be 'sold from the drawing board' and may arrive in the marketplace in a less than stable condition (one salesman's comment that 'this product runs on any overhead projector' typifies the market dynamics).

8.5 DESIGNING FOR MIGRATION

When installing the current enterprise network, a large amount of the design will actually revolve around how we migrate from the existing legacy network. In a similar fashion, today's network will become tomorrow's legacy—and if we have some strategy for the future shape of the network, we may be able to design today's network in such a way to facilitate the future migration. Some starter thoughts on migration strategies follow:

- Parallel running. It is likely that during the migration period, there will be a need for the new networking technologies to coexist with the old. If the network is a branch and hub type network, the new technology will first need to be introduced at the host site (little point migrating branches if they then have nothing to talk to!).

- Cost minimisation. It is desirable that cost of migration be kept as low as possible, and early savings from removal of legacy equipment are made as soon as possible. Often a key business case driver for moving to new technology is that licensing and maintenance costs of legacy systems are high. In hub and branch networks, this would tend to rule out parallel running for other than a short safety period, during which fallback to the old network is possible if the new one fails. As soon as possible after migration, the legacy equipment should be removed from the branch site.

- Regional migration. If a traditional SNA or TDM hierarchy is being replaced, there may well be a case for planning the migration in a regional fashion so that we can start to remove regional concentrators and their leased lines back to the hub sites at the earliest possible date.

- Technical preparedness. It may well be possible to anticipate the needs of evolving technology. Often new software versions that we may wish to use require more memory than today's version. If it is likely that software migration will be required, consideration should be given to supplying a larger amount of memory initially (revisiting all sites to add extra memory later will be an expensive and time-consuming activity). Again, we may be using Frame Relay today, but are looking at using ATM at a later stage—routers and other network components may be selected where the manufacturer has existing or planned capability to support the future technology.

- Backward compatibility. Upgrading a network can be made much more difficult if new versions of network hardware and software components are incompatible with today's versions. Where backward compatibility exists, networks can be upgraded in a gradual phased manner. Where there is no backward compatibility, a much riskier 'big bang' approach will be needed—and this approach will also invariably require a higher level of service affecting down time.

  Most manufactures do aim to ensure backward compatibility between one version and the next, but will make no such guarantee between today's version and code of two or more generations hence. It is therefore often necessary to periodically upgrade network software to keep up-to-date, even if new features offered are not required.

In addition to these issues, there are other complexities such as getting any important data that is resident on the old network onto (or, at least accessible from) the new one. A key point that holds across all of the above is to know what you actually have—a sound inventory of network components, users, data and services makes it a lot easier to plan ahead.

Table 8.1

Cause	Action	Issues
User data	Where possible, developers should work to minimise the number and size of messages that must be sent across the enterprise network. Data Compression should be considered. This is best done in end systems (e.g. files may be compressed before transmission), but may be done in routers, models or dedicated compression hardware. User guidance and discipline. An example would be the issue of large file attachments on e-mails. If the mail is to be sent to many people, it will be more efficient to store the attachment on a common file server and pass a file pointer in the e-mail.	If using routers to perform compression, this will place a significant processing load on the router. Care should be taken that this does not impact on data throughput. It is all too easy for departments to introduce new applications, protocols and LAN components without the network administrator being aware of it. This can have unforeseen impact on the operation and performance of the enterprise network. Regular network audit (using LAN analysers) is recommended, together with an efficient mechanism for handling user change requests.
Protocol Headers	As shown earlier, this can be a large source of overhead. Where possible, tune network to send user data in as large packets as possible. This reduces the percentage overhead of headers. Routers may be capable of performing header compression in many cases (much of a header does not change within the context of a session, or changes in a predictable way).	Making packets large can adversely affect performance of transaction oriented traffic and is best for file transfer traffic. Again, we need to check that the increased level of router processing is not an issue.
Unwanted Traffic	Careful use of traffic filtering using access lists can reduce load on networks.	
Broadcasts	'Broadcast Storms' can cripple networks. These occur when many sites have a need to broadcast data simultaneously (e.g. in a Source Route Bridged network, all workstations may explore for the host at the start of the working day). Routers are much more effective at suppressing broadcasts than bridges. Special features (such as 'proxy explorer'—where a bridge can cache paths to hosts, and respond to workstation explorers can help).	

Table 8.1 (*continued*)

Cause	Action	Issues
Routing Updates	The impact of Distance Vector Protocols such as RIP can be reduced by reducing the update frequency.	Increasing update interval adversely impacts network convergence time following faults.
	Use of a Link State protocol such as OSPF is more bandwidth friendly.	These protocols need careful design to avoid 'broadcast storm' problems following network faults as routers reconverge.
SAP broadcasts	This is an issue for Novell networks. Servers on these networks advertise their presence every 60 seconds, and this information is broadcast around the network. Routers should filter this information so that information about locally significant servers (e.g. print servers) is not broadcast.	
Management Traffic	The frequency with which routers are polled and the level of data collected must be carefully managed. It is easy to design a network that grinds to a halt every 15 minutes as the management system tries to read the complete MIB on every router in a network	There is a clear trade-off between responsiveness to network problems and the frequency of polling. Ability to forecast future growth requirements may be impaired if too little data is collected too infrequently.

8.6   MAKING THE MOST OF WHAT YOU HAVE

One final word on planning for the future is that you do not necessarily have to add new equipment. It is often worth seeing what can be done with what exists—the concept of network tuning. Through the use of this and other techniques, it is often possible to squeeze extra bandwidth out of an already busy network.

Table 8.1 shows some of the causes of network inefficiency, and describes what can be done about them:

8.7   CHECKLIST: HANDLING NETWORK CHANGE REQUESTS

In order to prevent network chaos, caused by uncontrolled change, a process that puts the following checks should be in place. It is assumed that a basic inventory is established beforehand as a baseline.

❑ User originates change request using standard format.

❑ Record change request.

❑ Assess and agree priority and required by date.

❑ Determine timing of change—can it be bundled with other changes and incorporated into a new network design version.

❑ Design and document technical solution for the change.

❑ Cost the change and agree this with the budget holder for the network.

❑ Ensure proposed changes are tested and known to work.

❑ Plan how change is to be implemented. If change affects many sites, consider implementing at one site first, then a small number a week later and then start full roll out a week after that.

❑ Ensure that there is a documented fallback plan if the change fails.

❑ Agree planned change timing with user representatives, and notify any users whose service will be changed.

❑ Carry out change.

❑ Test that changed network works as required.

❑ Ensure that all network documentation and management systems are updated to reflect the changes.

8.8   SUMMARY AND KEY ISSUES

This section has looked at how the enterprise network can evolve to meet the future needs of the business. We started by stressing the need to carefully control and manage change in the network. If control is lost and users start making ad hoc changes and additions, the enterprise network can very quickly become unstable, becoming part of the problem, rather than part of the solution to the business!

Given that change in the network is inevitable, we then looked at how we might proactively prepare and plan for that change, by forecasting and continual assessment of emerging technologies. Finally, we touched upon the need to carefully plan the migration from one generation of network to another, and looked at some of the ways this might be done.

Some of the key issues that may occur for the future are:

- It can be hard enough getting useful information on the traffic requirements of existing applications, let alone those that are yet to be written! Network capacity planning can therefore involve a lot of intelligent 'guestimating'.

- Crystal ball gazing on the futures of data networks is not easy—if you ask 100 networking experts you will doubtless get 100 different answers (and, if you are unlucky, quite a few more).

But for all that, change is inevitable and the better network is one with a measure of planned agility. The rate of change in communication will get faster—driven by technology, user demand and rapid organisational changes. So, planning for the future is not really optional, it is an essential part of staying in business.

BIBLIOGRAPHY

Davies, D., Sandbanks, C. and Rudge, A. (1993) *Telecommunications After AD2000*. Chapman and Hall.

Guilder, G. (1993) When bandwidth is free. *Wired*, September/October.

Naisbitt, J. (1994) *Global Paradox*. Nicholas Brealey.

Ohmae, K. (1990) *The Borderless World*. HarperCollins.

Porter, M. E. (1986) *Competition in Global Industries*. Harvard Business School Press.

Ward, J., Griffiths, P. and Whitmore, P. (1993) *Strategic Planning for Information Systems*. John Wiley & Sons.

9

Verification, Validation, Testing and Operation

We often find out what will *do by finding out what* will not *do*

Samuel Smiles

A significant part of the work that goes into building an enterprise network is creative thinking. For all the guidance and prescription that is introduced into the process, the quality of the end result still depends, to a large extent, on people with a good mix of experience and imagination. One of the lessons of experience is that ideas, proposals and approaches should be tested at all stages.

The traditional view of testing is that it is the activity immediately before the release of a finished product—the final series of checks for errors. Ideally, it should be associated with all stages of a design and should deal with the validation of requirements and the verification of specifications as well as the measurement and test of tangible product.

The idea that you need to 'check as you go' underpins much of the guidance in earlier chapters. The bonus in doing this is that it is a lot cheaper and easier to remove a problem at the paper stage of a design than it is to, for instance, reconfigure an installed and operational network! In the first part of this chapter, we explain what sort of methods can be used for verification and validation and just how you go about applying them.

No amount of testing will, in practice, guarantee a fault-free result. The complexity of modern networks means that there will always be some operational failures or unforeseen characteristics that result in corrective action having to be taken. Given this, the second part of this chapter moves on to explain how to cope with the (inevitable) problems that arise with operational networks.

9.1   VERIFICATION

It is essential that, during the design process, verification is carried out. This consists of a set of arguments, calculations and demonstrations that aim to show that the design will meet the customer's stated requirements. And the earliest point at which this can take place is when the customer states their initial set of requirements.

Conformance statement

Perhaps the simplest form of verification is the text-based Conformance Statement. Most contract bidding processes will ask tenderers to provide such a statement. In order to produce a Conformance Statement, the requirements need to have been clearly documented as a set of quantitative statements—ideally in the form of numbered clauses. A Conformance Statement is usually presented in tabular form, with columns as follows:

- The Requirement. The reference number of the requirement is given, and often, for the convenience of the reader, the requirement text will be reproduced.

- The Conformance Statement. This is usually a single word or phrase, e.g. conformant, non-conformant or partially conformant.

- Justification. This will be some text explaining why the design does or does not meet the requirement. Where there is non-conformance, a proposal should be given to address the issue (this could be through deferred implementation of the feature, or through a concession request, asking that the requirement be dropped or modified).

Risk assessment

Often, claims made in the Conformance Statement will be based to some degree on assumptions. This might be information assumed, based on experience, to cover gaps in the requirements. It may be an assumption that certain untried features of a router works correctly. Where risks exist, a record should be made for each risk, as to the likely impact on the network if worst fears are realised. Also, a statement should be given concerning any steps planned to minimise the risk (e.g. lab testing). Project administrators will want to see a financial impact assessment for the risk—how much to put the problem right and how much we might have to pay the customer in damages. Once all risks are documented, the probability of their occurence and the likely cost for each has been established, we can work out the sum of cost/probability products and get some estimate of the overall risk we are taking with the project. If the risks *must* be taken, the wise project administrator will build in a suitable allowance into the project pricing, in order that there remains a high probability that the project will remain profitable, even if some risks do materialise as real problems! While customers do not usually require the supplier to provide such a risk assessment, the tenderer would be well advised always to carry out this assessment, in order to understand and minimise the risks involved in taking on the contract.

Testing conformance to stated requirements does not come to an end when a bid is won. There is always some flux through the evolution of a design; customer needs change over time, some technical possibilities evaporate and others emerge. It is well worth investing a small amount of effort in maintaining the Conformance Statement. In addition to keeping an audit trail it can help with the assessment of network change requests. Similarly, there must be an ongoing commitment to maintaining the risk register. Results of testing and operational experience may eliminate or reduce the severity of some risks, whilst other previously unrecognised risks may come to light.

Calculations

A number of calculations will be required to back up claims made in the Conformance Statement. This section introduces some of the simple 'rule of thumb' methods commonly used by network designers when carrying out verification work. Readers requiring a more rigorous understanding of network design mathematics are referred to the specialists texts contained in the References.

Network delay or latency

Network round trip delay is a key factor in determining the quality of service (QOS) experienced by transaction oriented applications. Delay calculations will normally be required for mean and 95 percentile delay.

It is particularly important that a common understanding of the delay definition is agreed, including when the measurement starts and stops, what is included and what is not, and how long the messages are.

A good definition in the case of our Data Bank example might be: 'Round trip delay is defined as the time from when the first bit of a

128-octet application request message leaves the branch to the time that the last bit of a 2048-octet response message is received at the branch'. The calculation is to exclude all elements of branch end system delay (e.g. time to create the request packet and any queuing delay before it can be successfully be transmitted to the router). Also excluded are host end delays—measured from receipt of the last bit of the request at the FEP to the emission of the first bit of the response from the FEP. The calculation should include all protocol overheads added to the application message.

Figure 9.1 shows a typical network scenario and the important elements of delay that must be considered.

Serialisation delay

Because WAN speeds in many enterprise network designs are often significantly slower than LAN speeds, WAN access links are often the dominant cause of delay in networks. Data packets must be sent across the WAN access circuit one bit at a time, at the access circuit bit rate. Serialisation delay is easy to calculate:

Serialisation Delay = (Data block size in bits + Protocol overhead bits) / Access Circuit Speed.

By way of example, let us assume we are sending 1024 byte blocks of data, and each block has 40 bytes of TCP/IP header. There are 8 bits per byte. The Access Circuit Speed is 64 kbps.

$$\text{Delay} = (1024*8 + 40*8)/64000 = 133 \text{ milliseconds.}$$

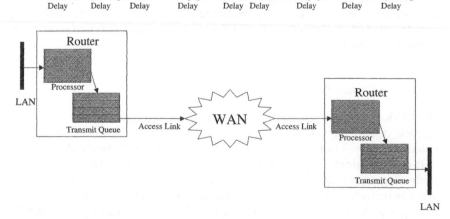

Figure 9.1 The Elements of Network Delay

Propagation delay

This element of delay is caused by the fact that signals travel at finite speed, and cannot exceed the speed of light in a vacuum (300 km per millisecond). In real network circuits, speed will be less than that of light in vaccuum, because (a) the cable will not take the shortest possible route, (b) signals propagate slower in cables or fibre optic than in vaccuum (c) repeaters and switching equipment add extra delay along the way. Generally, we will not know the exact cable routings and will only know the straight line distances between end points. In these cases, a good rule of thumb is to assume a propagation speed of 100 km per millisecond for land based routes. If a geostationary satellite is used, this will add a delay of some 260 ms.

Propagation delay becomes important (a) when working with large countries such as the USA, (b) international links—especially those using geostationary satellites and (c) where the application requires minimal delay to work properly.

Processing delay or latency

The processor in each network node (e.g. a switch or router) will take a finite amount of time to analyse each packet and determine how to onward route it. This (usually small) delay can often be found in manufacturer's literature or can be determined by laboratory measurements. (In some cases, only the packet per second throughput of a device may be known. If it is a simple device with a single processor, it is often safe to assume that packets are processed one at a time by the processor and therefore the reciprocal of the packet per second throughput is a good indication of processing delay.) Router processing delay is generally a very small component of the overall delay, typically being in the order of 100 microseconds to 10 milliseconds.

Queueing delay

In typical network designs, the speed of wide area network links are much lower than that of local area networks. If LAN users send a burst of packets toward a remote destination, the WAN access circuit will form a bottleneck. The packets will have to queue and await their turn. This is an especially important area of delay, because it is the one that introduces the greatest variability into the overall network delay. Some applications (especially those involving transport of real time voice or video information) are particularly sensitive to delay variations. Human users of interactive computer transactions also notice variable delay, and this can impair their productivity.

To estimate queuing delays, two models are commonly used, known as M/D/1 and M/M/1. In both cases the models assume that packets arrive randomly. The probability of a given number of packets arriving within a given time interval is assumed to follow a Poisson distribution. M/D/1 applies to networks and applications where the packet length is of fixed size. M/M/1 is for the more typical case with variable packet length—this model assumes a negative exponential distribution of size—i.e. small packets are most common, and the probability of finding larger packets decreases with size. Queuing delays calculated under M/M/1 represents a worst case scenario. Using this model tends to overestimate queuing delays and hence can add a useful safety margin into calculations.

In these models, a simple formula is available for calculating the average queuing delay (q), based on the average time to transmit (serialisation delay) each packet (s) and the utilisation of the link (u). Utilisation is the average amount of time that the data link is busy sending packets.

$$\text{For M/D/1, } q = (u{*}s)/(2{*}(1-u)) \quad \text{For M/M/1, } q = (u{*}s)/(1-u)$$

If a data link is 50% utilised, the queuing delay in the M/M/1 model is the same as the serialisation delay. Beyond 50%, the queuing delay starts to rise quite rapidly, due to the reciprocal element in these equations. That is why the network designer generally plans to keep network link utilisations well below 100%!

For these models, the probability distribution of delay against time follows a skewed distribution. The distribution has the property that the interval between no delay and the mean delay is one standard deviation. The 95 percentile can be estimated as being 2 standard deviations higher than the mean—i.e. 95 percentile = 3* mean delay. The 99 percentile can be estimated as being 3 standard deviations above the mean (or 4* mean delay).

Core delay

Network cores consist of transmission links, which introduce queuing, serialisation and propagation delays. They also consist of switching nodes which introduce processing delay. Therefore, the core delay has to be calculated as the sum of these delays, based on the expected route of traffic through the network.

Total end-to-end or round trip delay

Having determined the individual elements of delay, the total end-to-end or round trip delay can be calculated by summing the individual delay elements.

Delay variability and 95 percentile delay

Some elements of network delay are statistical in nature, with delays varying on a packet by packet basis. This is especially the case for queuing delays. It can also effect processing delays (as packets will have to queue for processing), but this is often a very small effect. When we discussed queuing delays above, we described how to estimate the mean and standard deviation of the delay for a single queue. To find the 95 percentile for a system with multiple queues, we need to find the standard deviation for the whole system. This is done by converting individual standard deviations into variances (by squaring the standard deviation) and summing the individual variances. The overall standard deviation is now given by the positive square root of the summed variances. The 95 percentile and 99 percentile can now be calculated as:

95 percentile = average delay + 2* Standard Deviation.

99 percentile - average delay + 3* Standard Deviation.

This is a simplistic approach, but the results produced do provide a fair indication of likely network behaviour. If more accurate results are required, the designer is recommended to use a network simulation tool.

When considering delay, it is especially important to consider the impact of one application or protocol on another. Supposing a branch is carrying out interactive ATM transactions to the IBM mainframe, and the branch manager decides to download a large report from a Novell central server—what impact will this have on the ATM transaction times? In a poorly designed network, the file transfer could cause the IBM SNA sessions to time out, and ATM transactions could fail altogether! Routers can help with this by prioritising traffic from one application type over another. Where techniques like this are used, simulation becomes the only effective way to reliably predict network delay performance under load.

Throughput

This is particularly important for file transfer applications. The customer will need a guarantee of the level of sustained file transfer throughput that can be maintained.

The primary limitation of throughput will be the speed of the slowest data link in the end-to-end path across the network. Even though an end system is connected to a 10 Mbps Ethernet, file transfer cannot exceed 64 kbps if that is the speed of access circuit used to connect the customer's site to the enterprise network.

To calculate throughput, we also need to understand the customer's end-to-end file transfer protocol. These protocols work by sending blocks of data and then waiting for acknowledgement. Generally, the protocol will allow more than one datablock to be sent without receipt of an acknowledgement. The maximum amount of data that can be sent without an acknowledgement is referred to as the window. We need to know the window size and the size of each block sent. It is common with window schemes for the end system to send a window's worth of data and then have to wait while this is propagated across the network. No more data can be sent until a block of data has been received at the far end and an acknowledgement has been sent and received. The following example in Figure 9.2 shows a simple example of a system having a window size of 2.

When window closure is the norm, throughput is heavily influenced by the round trip delay of the network. The earliest that an acknowledgement can be received in the above diagram is after block 1 has been fully received and the acknowledgement has propagated back to the sender. Under these circumstances:

Throughput = Lowest circuit speed * ((Time to send a window of data)/ (Time to receive first acknowledgement))

Ideally, the customer should tune their file transfer parameters to keep the window open. Note that the internet FTP protocol makes use of the TCP transport layer protocol. This features a 'sliding window' mechanism. When the TCP session is opened, it starts with a small window. As transfers are successfully acknowledged, TCP gradually increases the size of the window. This process continues until such time as the network starts to become congested. At this point, we can expect the network to start dropping packets. Once the TCP session has to start retransmitting, due to packet loss, it significantly reduces the size of the window, helping to prevent further congestion of the network. Designers should be aware of this behaviour. It means that Batch File transfer processes will attempt to use as much bandwidth as possible, and can severely impact any

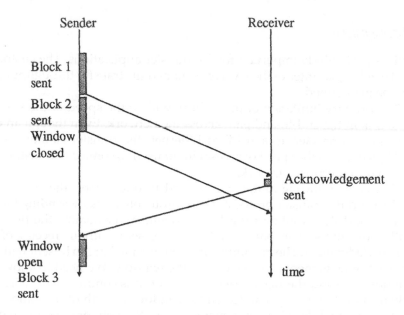

Figure 9.2 A Windowing System

transaction oriented traffic sharing the network. The answer is to use a traffic prioritisation technique to restrict the bandwidth available to the FTP protocol, ensuring that transaction oriented traffic gets a fair share.

In assessing the throughput, it is also important to know how many file transfer sessions at once are occurring at the host end. Most modern network designs use statistical multiplexing techniques, and the bandwidth at the host end is generally less than the sum of bandwidth at the branches. This is fine for transaction type traffic that is statistical in nature, but becomes the key bottleneck if a file transfer must be carried out to many branches simultaneously. The maximum throughput each branch will receive under these circumstances is:

Throughput = Host end bandwidth/number of branches requiring simultaneous transfer.

Impact of protocol overhead and line error rate on throughput

Throughput on data networks is often less than the above simple calculations suggest. To get a more accurate estimate, we need to take into account factors other than just bandwidth limitation and window closure.

We have already seen that TCP/IP adds significant overhead (40 bytes) to each block of user data. When we are calculating throughput, we are generally interested in throughput for useful end user data—so we will want to exclude the protocol overheads. This useful end user data througput Is sometimes refered to as 'Good-put'.

User data throughput = overall throughput * (user data block size)/
(user data block size + protocol overhead)

Real data networks are also exposed to line errors. Typically analogue links might have a bit error rate of one bit in every 100 000 on average. A digital link might corrupt 1 bit per million. When a packet is corrupted in the network, it will either be discarded because of a check sum error at the data link layer, or by check sum error within the TCP header. The end-to-end TCP protocol detects the loss, and causes the errored packet to be retransmitted—reducing useful throughput.

The probability of loss due to corruption within a packet is given by:

Packet loss probability = 1 − ((1 − bit error rate)**length of packet in bits)

To the probability of packet loss by line error, we need to add the probability of packet loss due to discard in the core network or routers. With TCP/IP, the standard approach to dealing with congestion in the network is simply to discard excess packets. The network supplier will design the core of a network so that congestion is minimised, and can be expected to publish target figures for packet loss in the network. The end-to-end TCP protocol will detect and recover from packet loss. This is done by resending packets. The more packets that have to be resent to recover from a lost packet, the worse will be the Good-put.

The following model offers an approach to estimating Good-put, taking into account packet loss through error or discard, resent packets and packet overhead:

Good-put = lowest link speed * (User bits per packet)/
((User bits per packet + Overhead bits)*(1 + packet loss probability * packets resent to recover))

Blocking

This calculation applies to a situation where client systems connect on demand to server systems. Examples are home workers dialling up to a fixed number of access ports, or branches with dial back up facilities dialling around the main network in case of local circuit failure to a fixed number of back up ports at the host.

In some cases, say where a client system will connect for long periods and must be guaranteed connectivity, it will be appropriate to dedicate a server port to a client. Generally, however, this is not cost-effective, and clients have to compete for a limited number of server ports. In this case, it is normal for a customer to agree a blocking rate, i.e. what percentage of calls at peak time will fail to find a free server.

The telephony industry has a useful model, known as Erlang B, which can be used for blocking calculations. First, determine the level of peak traffic that we expect to be offered to the network. This is measured in Erlangs and defined as the summed holding time for all calls, divided by the period over which the calls occurred—often described as call hours per hour. Given the traffic level and the number of server ports, one can calculate the blocking rate. Alternatively, given the blocking rate required, the number of server ports needed can be found. If T is the traffic in Erlangs and N is the number of ports,

Blocking Probability = $((T ** N)/N!)$/Sum for $X = 0$ to N of $((T**X)/X!)$

By way of an example, consider the case of a dial in facility for remote workers. Let us assume that traffic is handled by a remote access server (RAS), networked using ISDN primary rate service. The ISDN supplier offers best tarrifs if the network manager buys either 15 or 30 B channels configured on each circuit. The network designer may target the service for 2% blocking, but will accept 10% as a worst case before upgrade. The following graph in Figure 9.3 shows blocking probability given by the Erlang B formula against traffic for a RAS provided with 15, 30, 45 and 60 channels.

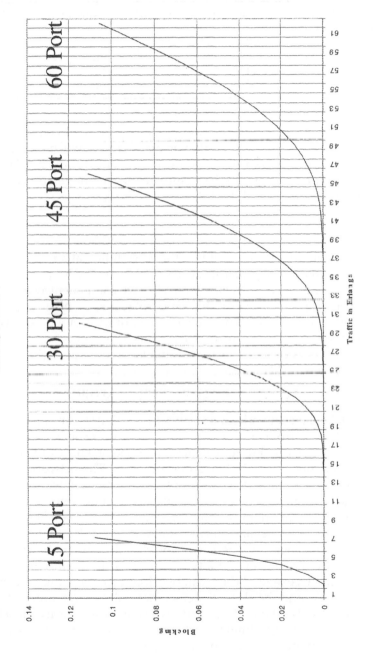

Figure 9.3 Blocking Probability Against Channel Numbers

Table 9.1

Number of B Channels	Users at 2% blocking	Users at 10% blocking
15	26	32
30	88	112
45	142	176
60	198	242

The network designer believes that 50% of users will connect to the system during the peak hour and that on average they will connect for 30 minutes. This means that each user is effectively generating 0.25 Erlangs of traffic. We can then use the graph to estimate the user population that can be supported displayed in Table 9.1

Availability

A typical definition for this would be the probability, at any time, that a branch site is able to communicate successfully with the host site.

This availability can be calculated given known component availabilities (e.g. for a router, a network connection, an ISDN line etc.). These components are arranged into an availability chain, representing the proposed network topology, and the aggregate availability is then calculated.

Availability is dependent on mean time between failure (MTBF) and mean time to repair (MTTR).

$$\text{Availability} = (\text{MTBF} - \text{MTTR})/\text{MTBF}$$

For well established network components, MBTF figures can be expected to have been derived from historical fault data. For new components, MTBF may have been derived through calculation, based on known availabilities of subcomponents. It may just be a design target. The network designer should always seek to understand the basis for any quoted availability figures, and form an opinion of how accurate they might prove to be!

MTTR is a very important customer measure, likely to be a key figure in the contract service level agreement. In agreeing availability levels, it is important to consider what period of the day this should be measured over—typically, a higher availability is required during office hours than at evenings and weekends. Where a network connection is mission critical and required around the clock, the customer will often require a 4-hour MTTR, provided 24 hours a day, 365 days a year. Less critical connections may only require fixing within 8 working hours (which frequently means the service is repaired on the next working day).

End-to-end availability for a path through an enterprise network will need to take into acount all the different components the data has to pass through, and any parallel resilience paths that have been provided.

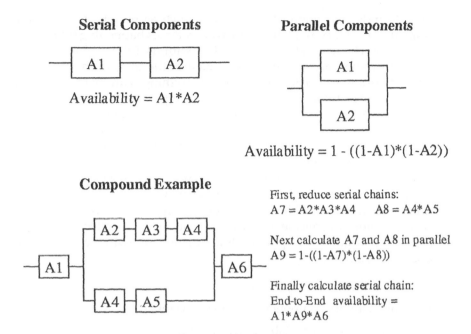

Serial Components

Availability = A1*A2

Parallel Components

Availability = 1 - ((1-A1)*(1-A2))

Compound Example

First, reduce serial chains:
A7 = A2*A3*A4   A8 = A4*A5

Next calculate A7 and A8 in parallel
A9 = 1-((1-A7)*(1-A8))

Finally calculate serial chain:
End-to-End availability =
A1*A9*A6

Figure 9.4   Combined Component Availabilities

Component availabilities are combined as shown in Figure 9.4. A worked example, using typical availability figures can be found in section 10.5.

The network designer needs to be aware that component failure is a random process, and that availability calculations are based on a statistical process. The predictive value of availability calculations is therefore at its best when dealing with large networks.

9.2   VALIDATION

If verification is, broadly, about checking that you are doing it the right way (at least in theory), validation is about making sure that the theory translates into practice as you would want. In many ways, this is a matter of being conversant with the practicalities of the design elements being used and with the real-world issues that arise in systems development and network operations.

Components

With today's rapid pace of technology change, it is certain that you will be using components in the design of the enterprise network that are new (or even still under development!). Products, such as routers, are also in a continual state of flux, with major software releases happening two to four times a year. The new releases are packed with useful new features (one reason why routers are sometimes referred to as the 'Swiss army chainsaw' of the network world. You can do almost anything with them, including serious damage, and fast, too!), some of which you may have assumed in the network design (e.g. the latest release may offer to support Bisync SNA as a legacy protocol). It is a sad fact of life that first releases of such new features often contain bugs, so testing of any new feature you intend to use is highly recommended.

It is inevitable that, during top level design work, a number of assumptions will have had to have been made about the operation and performance of components. It is a good idea to test these assumptions at this stage also.

Before committing business-critical applications to such technologies, it is essential that small-scale pilot trials are carried out with the technology. These should aim to:

- prove that the technology works;
- integrate network components with end systems;
- show that it offers acceptable performance;
- demonstrate reliability;
- gain early exposure/training on new technologies;
- gather data required for implementation design;
- validate design assumptions.

When a customer deals with an outsourcing network supplier, it is common practice to negotiate a trial of networks and equipment before entering into full supply contracts. Such trials will be carried out at this time.

Testbeds

It is recommended that any company embarking on implementation should establish a permanent testbed. It is likely that the company will already have a test host and test branch locations which will be used for application development. It is now desirable that representative network connections and the associated routers are added to this testbed, so that the network can be integrated with the customer's end systems and applications.

Having laid plans to set up the testbed, the next step will be to design a set of integration and acceptance tests that can be used to prove that the customer's applications will satisfactorily inter-operate over the proposed network. The design process is also likely to have suggested a set of tests to verify assumptions concerning functionality and performance of networking components.

All tests should have well-defined success criteria. This is especially important when tests form part of a trial contact with a supplier. A suitable set of test success criteria will need to be agreed with the supplier and included in any trial contract.

When the testing is performed, it is important to ensure that results from tests are well documented, and quantitative information obtained is fed into the detailed design process.

Integration

Before end-to-end application testing can start, it is necessary to integrate the components of the testbed. Having installed and interconnected the network components, they must now be configured, and basic connectivity checks carried out.

It is strongly recommended that a step by step approach to integration is taken. If we try to get the whole system operational in one go, it can be difficult and time consuming to track down problems that may occur. A suggested approach is:

- The network provider should configure hub routers and their wide area connections. Confirm connectivity (e.g. by using the TCP/IP ping facility).

- The network provider now connects and configures host end access/ encapsulation routers, again confirming connectivity.

- The customer's IT staff now work in conjunction with the network provider engineers to attach branch and host end systems, configure them and carry out end-to-end application tests.

The output from the logical design step should have provided adequate guidelines for configuration of the testbed components, but this document should now be revised to reflect any lessons learnt during testing.

During testing, it is certain that problems will come to light, and essential that a disciplined approach to fault logging and tracking be developed. Useful information can be gained at this stage about a supplier's responsiveness and competence at resolving system and integration problems.

9.3 TESTING AND DEMONSTRATIONS

The importance of testing cannot be stressed enough—even the best plans, implemented in the right way stand a chance of coming to grief. The designer would do well to adopt as a motto 'If it hasn't been tested, it doesn't work!'. We need to consider testing of all new components in a design (and with the pace of change with component software in devices like routers, even established models should be treated as 'new'). Most importantly, we need to consider integration testing, including a customer's end systems with routers, routers with networks, and full end-to-end application testing of the network.

As a minimum, plans should be laid at this stage as to what must be tested at the trial stage of the product (see later sections for more detail on testing). If any particularly new/risky design approaches are being proposed, it is as well to start lab testing immediately, however.

Some customers will insist on seeing a demonstration of the proposed solution as part of the tender evaluation process. A common minimum requirement is a visit to the network management facilities that will run the customer's network.

As well as testing functionality and performance of the proposed network, a key role of the testbed will be to assist in the tuning of the network to optimise performance. It will not always be possible to determine the optimum settings for the many router and end-system configuration parameters during the design process, and they may need to be determined experimentally, or tuned, in the lab. Areas where tuning might be required include:

- window and packet size

- prioritisation

- routing protocol parameters

- traffic buffer allocations.

With the complexity of modern networks, it is not practical to carry out exhaustive testing—there are simply too many aspects to cover. A good test plan has to trade off the time and resource available with the level of testing that should be completed. Experience shows that there is a diminishing return from tests, with a lot of errors picked up quite quickly and fewer emerging as testing proceeds. A clear test plan that identifies priority features and cases can be generated though the validation and verification stages.

9.4 KEY ISSUES IN VV&T

A key point to reinforce about validation, verification and testing is that it is never complete. The whole idea of making checks throughout the design and delivery process is to reduce risk of failure. Experience plays a large part in knowing where to focus limited time and resource. Some of the major issues are:

- Special attention must be paid to the testing of legacy systems. Support of these via modern LAN interconnect technology can be complex, and is not without compromise. Testing must be used to ensure performance will be adequate, and to allow for any necessary configuration tuning.

- There may be difficulties in scheduling the time and resources (both the technical staff and equipment) needed for this testing. However, effort spent on identifying and resolving problems at this stage will be more than repaid. Just think of the hassle there would be if a major design issue came to light halfway through operational roll out!

- Obtaining suitable test equipment may be a problem. Hiring may be a suitable approach, and may help with decisions about what test equipment to purchase to use in supporting the operational network.

- It is essential to test important combinations of protocols and applications. This is because they may interact in an unfavourable way. A typical example is where one user is carrying out transactions with a remote host, and a second user starts a file transfer. It is possible that during the file transfer, the transaction application may slow down, giving unacceptable response times—at worst, the application session could time out and the user is disconnected. If such problems are found, then work is required on the router configurations, bringing prioritisation mechanisms into play.

- Flood testing is to be recommended. In this, the aim is to stress network links, routers etc., with sustained high levels of simulated traffic, up to the peak levels expected in live use. Many types of networking problems only come to light under such load conditions. It would be clearly very embarrassing if the problems come to light for the first time when the network is heavily loaded with live traffic. Generating realistic traffic at the required levels can be a problem. Options are to:

  - use a number of high performance LAN analysers as traffic generators/sinks;
  - develop variants of branch systems which can be set up to generate high volumes of simulated traffic;
  - traffic may be generated from a traffic simulator running on a mainframe. FEP ports can be connected into the enterprise network testbed to simulate concentrated branch traffic.

At this point, we move from having a captive network to having one that is installed with the customer and running live applications. Even if it was perfect when it leaves our hands, there will be teething problems once it goes live. So we have to keep it going, and this means that some thought will have to have gone into problem resolution post release.

9.5 PROBLEM RESOLUTION

The technology used in enterprise networks is evolving rapidly, and many of the systems being used will be relatively immature. It is almost inevitable that the user will experience some significant problems with the network, especially in its early days. The network designer is frequently drawn in to help with the investigation of major network incidents. There may also be lessons to learn concerning the adequacy of the design.

In addition to direct involvement with fault finding, the network designer will also be involved in designing the management systems and processes. These need to become increasingly proactive in the detection, and where possible, automatic correction, of network faults.

Identifying the problem—reactive v. proactive

Network problems become apparent to the users when either their applications fail to communicate, or operate with degraded performance. Once the user has noticed the problem, they will report it to the Network Management Centre.

This is reactive fault finding. This approach is seldom satisfactory for today's business-critical applications however, since the user will be looking for proactive fault finding. This is where network management systems are able to detect the failure before the user, allowing the network manager to start fault resolution before the user reports the problem. (Ideally of course, the network design will automatically self-heal after many faults, the management systems then indicate the location of the fault to allow repair.) Some techniques for proactive management (e.g. router polling, benchmarking) have been described previously.

Assessing the impact and logging the problem

First instincts will be to attempt to remedy the problem as quickly as possible. When dealing with a large enterprise network, where multiple faults are often in hand simultaneously, a more disciplined approach is required. First priority will be to log the fault and determine its priority.

The network operator should be supported by an automated fault logging tool. This will allow the fault to be logged into a database, together with subsequent progress reports and fault clearance details. These systems should provide automatic progress prompting, and escalation if the fault is not being resolved within the agreed time. Statistics on fault rates and repair times generated from these systems are essential evidence for demonstrating that Service Level Agreements are being met. When logging the fault, the network operator will:

- Capture as much information as possible about the symptoms of the fault, and the circumstances preceding it.

- Notify the customer (if the fault was found proactively) and provide a fault reference number.

- Determine a severity level—this will influence the priority with which the problem is dealt with, and often determines the contractual time allowed to fix the problem.

The severity level scheme will be agreed between the customer and the network operator. For example:

FATAL—A customer site has lost all service.

SEVERE—A site has service, but with degraded performance.
MILD—A site has a fault, but service is unaffected (e.g. a site has
        successfully resorted to an ISDN back up path.

Of course, some faults do not occur on a per site basis, but may affect a whole region of sites, e.g. a Frame Relay node fails. It can be helpful to keep an overall status level on a given customer's network:

RED—greater than 5% of sites have service problems.
AMBER—greater than 0.5% and less than 5% have service problems.
GREEN—less than 0.5% of sites have service problems.

These figures are indicative only—the real point is there should be some quantitative way of assessing the health of the network. An approximate answer to the key question is preferable to an exact answer to the wrong one.

Gathering the evidence—first catch your bug!

Many of the faults that occur in a network are routine in nature, e.g. failure of a circuit or hardware failure of a router. It is reasonable to expect network operations staff to resolve such problems without recourse to the designer. The designer would be wise to examine network failure statistics on a regular basis. A useful tool here is 'Pareto Analysis'. This involves plotting a fault cause distribution, ranked in order of decreasing frequency (see Figure 9.5).

It is commonly found that some 80% of problems are due to only 20% of causes. The network designer can most profitably focus attention on resolving these. In doing this analysis, one can choose to look at either the

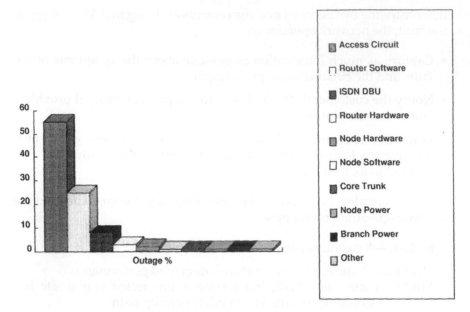

Figure 9.5 Pareto Analysis of Outage Causes

frequency of the fault type, or look at the amount of outage (lost service time) caused by that fault type. The figure analyses faults by outage. In this case the first two columns (Access circuits and Router software) are responsible for 80% of the outage, and most effort should be focused on these aspects to achieve radical service improvement.

Access circuit issues could be tackled by working with the supplier to analyse root causes of these failures and correcting them, or insisting that the supplier repairs faults more rapidly. Alternatively, the designer may recommend more widespread deployment of resilience techniques such as ISDN back up to overcome the weakness.

The second most frequent cause in this example is router software problems. A tactical fix here would be to see what could be done with the management systems to detect the failure faster, and to proactively re-boot the router. For a longer term fix, urgent work is required to find the root cause of these failures and repair it. Techniques for this are covered in the remainder of the chapter.

The designer is most likely to be called in when the network behaves in unexpected ways, due to software problems with routers or network switches. It is the analysis and resolution of this class of problem that is the primary concern of the rest of this chapter.

It can prove difficult to find the underlying cause of problems. Any 'easy' problems are likely to have been uncovered and corrected during network testing. This will leave faults that occur only intermittently, or under a very specific set of circumstances. It is also not uncommon for the network operations team to resort to first aid remedies, such as power cycling or re-booting nodes or routers. This will often resolve these problems, but in doing so, any useful diagnostic information available from the component is lost.

Pinning down such problems can require a great deal of patience and intellectual effort. The following may help:

- Which sites were affected—is there any relationship between them (e.g. the affected sites may have all been installed recently, or they may all have connections routed over a common trunk circuit in the core network.

- Try to find out as much as possible about what the applications were doing before the failure, e.g. does the fault only occur during high levels of traffic experienced during overnight batch file transfers?

- What were network conditions like before the failure?

- Is anything unusual to be found in the alarm log?

- If fault occurs repeatedly, is there a pattern (time of day, when the network is busy, etc.).

Having gained a working understanding of the problem, it is well worth asking 'Has anyone ever experienced this problem before?'. Component manufacturers frequently maintain databases of known problems. These can be searched on-line (in the case of the Cisco router, users have access to the Bug Navigator, as part of the Cisco CIO on-line information service).

If we are lucky, someone may have reported our problem before—there may even be a recommended fix (use a new version of code, load this patch, change the configuration, etc.). Failing that, there may be news of future plans to resolve the problem.

Frequently, the evidence available at this stage will not be sufficient to diagnose the problem. The best hope is that it will tell us where and when to look for more evidence. This will often involve sending technical staff to sites likely to present the problem. Evidence can be gathered by:

- Using Datascopes to probe the LAN and/or wide area network side of the site router;

- enabling diagnostic debug traces on routers and capturing the reports (this information is frequently requested by the router manufacturer when they become involved in trouble shooting. They will indicate precisely which debugging codes should be run to help them understand the issue).

Reproducing the fault

An ideal situation exists where the mechanism of the fault becomes sufficiently well understood that it can be reliably reproduced in the testbed which we constructed earlier.

Where the fault can be reproduced, effort can now be focused on testing further to establish exactly which network component or components cause the fault. We can also investigate the cause of the fault in a more vigorous way than is often the case in a live network. In the live network, the customer is likely to be sufficiently upset that the problem exists in the first place, without the network supplier making things worse by repeatedly attempting to reproduce the same fault for diagnostic purposes!

Resolving the problem

Obviously, to resolve the fault we must pin it down to one or more network components which we believe are to blame for the fault. We expect that changes to these components will resolve the problem.

A major difficulty can be getting the component manufacturer to accept responsibility for investigating and correcting a problem. The best way around this is to take the effort to build an effective working relationship with suppliers long before faults start occurring. This is perhaps easier for the larger customers of such components (an advantage that network outsourcing companies often enjoy). Some of the reasons why it can be hard to get a component manufacturer to accept responsibility are:

- 'You are not covered.' Component manufacturers will often price their software maintenance and support programmes as a separate optional item. It may be preferable, as a means of reducing network costs, to decline this option, but this may prove a false economy!

- 'You have not given us enough evidence.' This is one of the reasons why reproducing the problem in the lab is so important. If this is not possible, get the supplier to provide details of what they do need, and of how to get it.

- 'Our box is fine—the problem is with the system you have connected it to.' This is a commonplace problem when working with multi-vendor networks. Problems of incompatible interfaces are commonplace. Of course, the interfaces between systems will conform to published standards, and without these, interconnectivity would not be possible at all. However, standards are far from perfect—some of the issues are:

  - They evolve. New standards are often full of statements to the effect 'for further study', or exist in several competing versions. Manufacturers are keen to launch products using new standards, often doing so before the standard is fully complete and agreed. Different manufacturers will make differing interpretations of draft standards. If your network is using 'leading bleeding edge' technology, insist on careful lab integration testing of components. If the problem comes to light in a live network, you may have to migrate both systems to new versions of code which implement firmer versions of standards. Otherwise, you will have to negotiate for software patches to fix the problems. It pays to maintain knowledge and expertise of evolving standards in-house wherever possible, so that the risk of implementing new standards can be better assessed.

  - Standards are open to interpretation. They are generally written in English and are therefore open to ambiguous interpretation. There is a trend to increasingly use of logic based formal methods to define standards, which will help in future.
  - Standards are often full of options. Different manufacturers may choose different subsets. The standards community uses the concept of 'profiles' of standards to agree fixed subsets of options. New standards may not yet have been profiled in this way.
  - Manufacturers may decide to enhance a standard with proprietary extensions. These may affect interworking with other manufacturers equipment.

• 'Our box is fine—it's the way you are using it.' The supplier may argue that you have the system configured in an unsupported way, you do not have enough memory or you need a more powerful component. These problems, if true, can only be eliminated by sound design and testing in the first place. This is somewhat naive, as the designer is unlikely to have the same level of performance information or insight into the inner workings of the component as the supplier has. It is best if the supplier can be involved in producing or reviewing the original design. In these cases, it may be necessary to test the suggested design enhancements in the lab to see if it does resolve the problem. It then becomes a question of how the required network upgrade can be financed!

Whatever the difficulties encountered, it is important that you keep track of efforts to resolve the problem. There are many products with a 'trouble ticketing' capability which allows actions taken, escalations etc. to be tracked. With ever more complex chains of dependency between carriers, supplier and agents, more often than not it is worthwhile automating the problem resolution activity.

Implementing the solution

The solution, which usually involves a software upgrade, is handled much the same as any other network upgrade. It is implemented progressively after careful testing. There may be pressure to rush a 'fix' into the network rapidly—this should be resisted, as changed software versions can introduce new problems!

Once a solution has been identified, it is then a question of thoroughly testing it in the lab. In many cases, it is logically impossible to show that the problem has been fixed absolutely (it would take an *infinite* amount of time to demonstrate that a system *never* exhibits a given behaviour!). However, if we are able to reproduce the fault in the lab, we can show that the fault does not occur in these same circumstances with the fix in place.

Suppose we were never able to reproduce the fault, but a fix has been proposed? In this case, the best we can hope for is a reasoned explanation of what was causing the problem and why the manufacturer thinks the proposed fix makes a difference.

It is probable that the customer will want to run through a reasonably large set of 'regression' tests, to ensure that the supposed fix does not adversely affect the basic end system and application support functionality of the network.

Once lab testing is complete, the changes will be rolled out to a small sample of sites, and the impact on these sites will be assessed. If all is well, the fix can now be rolled out to all affected sites.

9.6 CHECKLIST: GATHERING EVIDENCE

When investigating network faults, the following checklist may be of use:

❏ How did the problem present itself—what were the symptoms and how was it detected?

❏ How is customer service affected—how many sites down and for how long?

❏ When did the problem occur?

❏ What was happening in the network at the time? What were the customer's applications doing at the time?

❏ Where did it occur—which sites were affected? Is there any connection between the sites, e.g. shared path through network or all installed at a similar time?

❏ What information is available from network alarm logs?

❏ What 'First Aid' remedies have been used to restore service—can this method be used tactically for any repeat occurrences? Does this remedy suggest anything about the root cause of the problem, e.g. if re-booting the router fixes the problem—could it be a router software problem?

❑ Have you full and up-to-date documentation of the network to hand? (You will require full network maps and details of the logical design—especially configuration of routing and protocol support.)

9.7 SUMMARY AND KEY ISSUES

This section has explained what sort of checks the designer can and should do as a design evolves in order to make sure that it is fit for purpose. This process of validation, verification and testing can begin at a very early stage, as the customer requirements are being put together. Methods for validating requirements and verifying design options are explained in this chapter.

Once a design is complete, checking that all is well moves from being a logical activity to a physical one. Testing is never complete as any network evolves (both through intent and accident) as it is used. So mechanisms for problem and issue management need to be put in place to handle this. There are reasons why and times when a designer may get involved in network troubleshooting. This is when the network is exhibiting unexpected behaviour that cannot be explained by any known failure mechanism.

A methodical approach to investigating the problem has been explained. This involves the following stages:

• gathering evidence

• attempting to reproduce the fault

• negotiating with equipment suppliers for a fix

• testing the fix in the lab

• progressive roll out of fix.

Although simple in principle, the maintenance of an operational network is a demanding task. Key issues with network troubleshooting discussed in this chapter are:

• Enterprise networks often incorporate state-of-the-art components. These are frequently flawed. Careful lab testing before network roll out can prevent many, but not all, live network problems.

• Reproducing and understanding faults in complex modern networks can be difficult and time-consuming, more art than science. People skilled in this area are rare and expensive, so is the equipment they need to tackle the job.

• It is not unusual for complex modern networks to occasionally experience a problem, which subsequently is never seen again. A great deal of time can be expended trying to explain these, often with access to inadequate data. Network problems often only become tractable to investigation when a pattern emerges.

* Many network problems are actually found to have multiple causes. When investigating problems of this kind, it is recommended that you try to fix one thing at a time. Trying to change too much at once can add to the confusion!

One final thing to say is DON'T PANIC. When faults occur, there is a tendency for both customer and supplier to panic whereupon the problem gets escalated to senior management levels. This can have a negative effect on the people trying to fix the problem (who can end up spending more time in meetings or filing progress reports than actually working on resolving the problem). It is well worth investing management effort to ensure the fault team have the resources they need, and work to keep customer/supplier relationships as calm as possible. Many complex network problems can be resolved by calm intellectual reflection and the right environment must be created to allow this to happen.

BIBLIOGRAPHY

Davis, A. (1990) *Software Requirements Analysis and Specification.* Prentice-Hall.

Gelenbe, E. and Pujolle, G. (1998) *Introduction to Queuing Networks.* John Wiley & Sons.

Grady, R. and Caswell, D. (1987) *Software Metrics—Establishing a Company Wide Program.* Prentice-Hall.

Harrison, P. and Patel, N. (1993) *Performance Modelling of Communication Networks and Computer Architectures.* Addison Wesley.

Hetzel, W. (1988) *The Complete Guide to Software Testing.* QED Information Sciences.

Kershenbaum, A. (1993) *Telecommunications Network Design Algorithms.* McGraw Hill.

Leintz, B. and Swanson, E. (1980) *Software Maintenance Management.* Addison-Wesley.

Musa, J., Iannino, A. and Okumoto, K. (1987) *Software Reliability—Measurement, Prediction, Application.* McGraw-Hill.

Marciniak, J. and Reifer, P. (1991) *Software Acquisition Management.* John Wiley & Sons.

Norris, M. (1995) *Survival in the Software Jungle.* Artech House.

10

A Case Study—Data Bank

In my course I have known and, according to my measure, have co-operated with many great men: and I have never yet seen any plan which has not been mended by the observations of those who were much inferior in understanding to the person who took the lead in business.

Edmund Burke

It is one thing to explain a way of doing things, another to apply it. In order to illustrate the design methods described in this book, we now present a major worked example, called Data Bank. Confidentiality agreements prevent us from describing a specific real customer, so the example given is fictitious, based on the authors' experiences across several projects the financial services industry. The small case study used in Chapter 2 was theoretical, as evidenced by the fairly clinical nature of the requirements. Here we see a more realistic, often vague and very diverse set of requirements.

The key aim is to illustrate some of the real-life complexity of enterprise networks and how this can be tackled. At each stage of the design process there are major choices to be made. It would clearly be an impossible task to describe in depth every combination of choices. But at each stage of the design process, an appropriate choice will be made and explained. The choices made will help to constrain and focus the discussion.

10.1 INTRODUCING DATA BANK

Data Bank is a UK based bank, whose principal business is high street retail banking. It has some 400 branches located around the UK. Data

Bank has, through an acquisition, recently moved into foreign exchange and securities dealing, and has a small, but growing, number of overseas offices.

Data Bank management realise that, in order to remain competitive, they must invest in a modern enterprise network. Of course, Data Bank already has an extensive network, and this is introduced below.

10.2   THE NETWORK TODAY

Data Bank is a typical 'True Blue' IBM house, with an extensive SNA network. The current network is summarised in Figure 10.1. The diagram illustrates a typical SNA hierarchy with terminals and ATMs (automated teller machines, or cash points) at the branches which are connected to a cluster controller. This is connected to a regional front end processor, located at a regional office. The regional FEP is then connected to the host FEPs via the FEP network, and hence to the mainframe hosts.

Host sites

It will come as no surprise that availability of the banks host computer is mission critical. No transaction can take place without reference to customer records held on the mainframe. To help ensure high host availability, Data Bank operate two similarly equipped host sites, located 100 miles apart. The host sites operate on a worker/host standby basis. This means that at any time, only one of the hosts is serving the branch network. The other host will be receiving database updates from the working host (a process known as shadowing), and will be ready to take over from the other host should a failure occur there. Figure 10.2 shows the host sites in more detail.

Each host site centres around an IBM 3090 mainframe. This communicates with the world via its IBM 3745 front end processors (FEPs), which are connected to the host using the proprietary IBM Channel Attachment technology. Each site has two FEPs. The two host sites are interconnected by a pair of dedicated 2 Mbps links, which are used for database shadowing and inter-site file transfers. Each regional FEP will have a connection to one of the FEPs at each host site to ensure resilience. There are 16 regional FEP sites. Clearly, no bank operates in isolation, and Data Bank relies on extensive links to other banks and inter-bank networks such as VISA. All such links terminate at the FEPs.

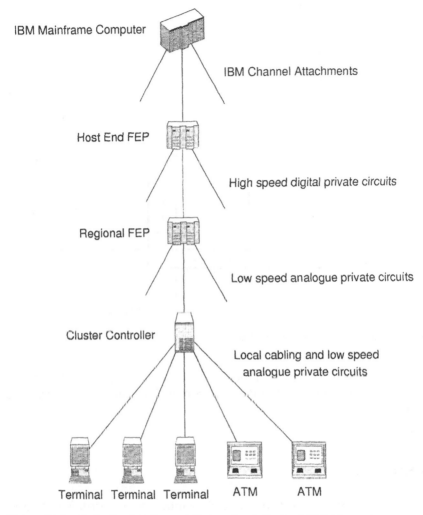

IBM Mainframe Computer

IBM Channel Attachments

Host End FEP

High speed digital private circuits

Regional FEP

Low speed analogue private circuits

Cluster Controller

Local cabling and low speed analogue private circuits

Terminal Terminal Terminal   ATM   ATM

Figure 10.1 Data Bank's Current Network—A Typical SNA Hierarchy

FEP network

In order to interconnect the FEPs, Data Bank take 64 kbps channels from a TDM backbone network, which the bank owns and operates. The majority of bandwidth on this network is used for voice traffic, interconnecting PABX switches at the banks main sites. Host sites and each of the 16 regional offices have a TDM multiplexer, to which the FEPs are connected. The TDM multiplexers are connected together in a partially meshed network, using leased 2 Mbps E1 leased circuits. The TDM multiplexers

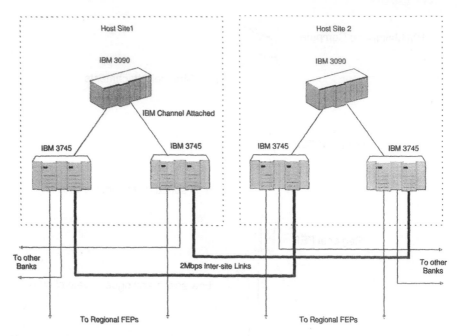

Figure 10.2 Data Bank's Host Sites

are meshed in such a way that there is always spare capacity, via a diverse route, to replace any failed FEP to FEP connection. This is shown in Figure 10.3.

Regional offices

There are 16 of these offices located around the country. The regional office in London is also the bank's headquarters building. Regional offices contain administrative offices (e.g. personnel, pay and training groups). These make heavy use of PC based office automation. In some departments, these PCs have been linked using Token Ring LANs, allowing shared access to file servers, printers and Token Ring attached IBM cluster controllers, for access to applications on the banks mainframe. In other departments, PCs are standalone, or attached only to IBM cluster controllers via coax cabling. A range of different OA software is in use across the company.

The regional offices provide some level of back office support for branches (processing of cheques, typing pool for the bank managers etc.). An IBM AS400 system is used for local processing in this role. This is able to communicate with the mainframes at the bank's host sites.

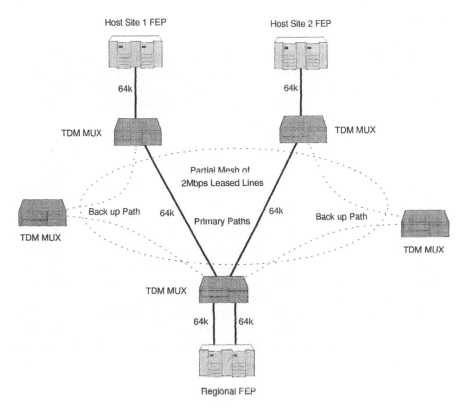

Figure 10.3 FEP Interconnection Network

Recently, the bank has been running an initiative with all its senior managers, called ManageNet. This has seen them all supplied with high end PCs and a common OA software suite. All ManageNet users have access to a common file server and e-mail platform. This has been achieved in a fairly ad hoc manner, using LAN bridging technologies. ManageNet has been developed using Novell 4.1 Network Operating System on its file servers and Windows 95 on end user client PCs, running Novell Client software. ManageNet also includes an intranet facility. This consists of a Web server running on a PC based server, running a Unix operating system. This holds pages of key Data Bank documentation, such as management procedures, descriptions of the product range and also a staff directory of senior managers (including photographs of them!). Client PCs run a Web browser to access this information. The Novell network used the IPX protocol in the network, whilst the intranet system relies on the TCP/IP protocol suite. ManageNet has improved the flow of information in the company at the senior level, and has generally been judged a great success. In the light of this, Data Bank now wishes to roll out this technology to all branch managers and their secretarial support staff.

Branches

The bank likes to equip each branch in a standard way. This gives the customer a consistent feeling of service, and makes setting up of new branches and staff training easier. The bank aims to have a maximum of 400 branches located around the UK. Of these branches, some 100 are located in cities, and have peak traffic loads typically twice as high as branches in towns. Figure 10.4 shows the layout of a typical branch.

Branches have three customer counter positions, each with an IBM terminal. The branch also has two ATMs. The manager's office contains an IBM terminal and his secretary has an IBM PC, which is able to emulate an IBM terminal and also needs mainframe connection. There is a further terminal in a customer interview room, and two terminals in the bank's 'back office', where cheques are processed, money counted and coffee consumed! All branch IT systems are wired to an IBM cluster controller. This cluster controller operates using the IBM SDLC protocol. This can support multidropped terminals, where a group of terminals can share cabling and a common port of the cluster controller (the controller controls use of the shared cable, communicating with each terminal in turn, passing it any received data and inviting it to send any data it wishes to transmit. This process is known as polling). Counter terminals, ATMs, etc. are connected in this way.

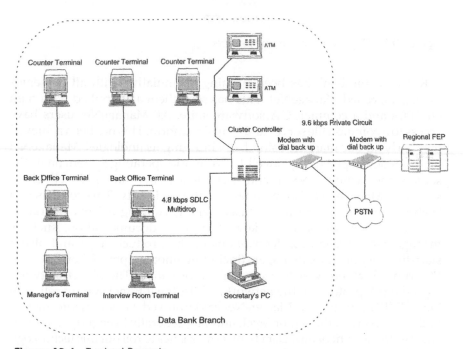

Figure 10.4 Typical Branch

The cluster controller has a leased line connection, operating at 9600 bps, to the nearest regional FEP. To guard against failure of the leased line (which tends to be the least reliable component in any communications network) the modems at the branch and FEP port are connected to the public telephone network, and can dial around the failed leased line. Voice traffic from branches is handled via the public telephone network— small branches have two lines and large branches have four.

International dealing network

Data Bank's currency and securities dealing operation has offices in four overseas locations (New York, Tokyo, Paris, Hong Kong). The headquarters and main host site is in London. The operation is successful and ambitious, with plans to add two new overseas offices per year (Singapore and Frankfurt are expected next). This is summarised in Figure 10.5.

As previously mentioned, this operation was acquired from another bank, and the communications infrastructure is entirely different. The system is based on client–server architecture. Overseas locations have client PCs attached to Ethernet LANs and the London site has large database servers, with which these clients communicate for transaction

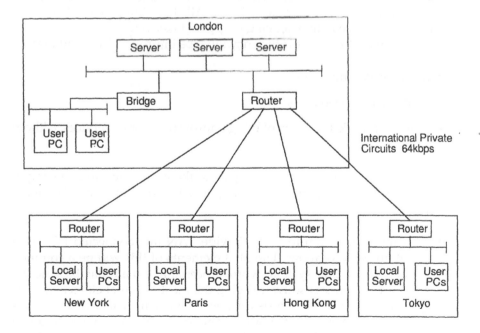

Figure 10.5   The International Dealing Network

traffic, together with file servers and e-mail post office server. The overseas offices also have local servers for file storage and print sharing. All software is Microsoft NT based, and the TCP/IP protocol suite is used. The remote sites are hubbed into London using routers, connected by 64 kbps digital International Private Circuits. These links have proved expensive, and reliability has been a problem, leading to loss of business at some locations.

The London site also has client PCs. These are connected to their own LAN, which is linked to the server LAN by a bridge. The bridge prevents server traffic from overseas locations appearing on the London client LAN. This helps the London client LAN to maintain a high performance with its local office automation (file and print serving) duties.

10.3   THE DREAM

The board of Data Bank know that they are one of the least efficient banks in the industry. They are also relatively small in size compared to the major UK high street banks. City analysts will compare different banks efficiencies using simple metrics such as customers per head of staff and turnover per head of staff. The impact of low efficiency is that charges to customers have to be relatively high to cover the banks costs—and this leads to loss of market share. If the bank reduces charges to retain customers, profits and hence share value will fall. All these things have led to the bank's share price falling, leaving it ripe for takeover!

If the bank is to survive, efficiency must be improved. This will mean:

- reducing staff numbers

- reducing overhead costs

- increasing market share by offering improved customer services, without additional staff.

The board believes that at least part of the answer lies in investing in improving the information technology used in the company. The main objectives are:

- LAN interconnection for OA. Provide LAN interconnected PC based OA at all bank locations. Essential elements of this will be a common package (operating system, word processing, e-mail, database access and IBM host access software being the most essential elements). This system will reduce staff costs (managers and staff will have less access to clerical support at branch level). The bank will also make substantial savings on its postal bills—most contact between branches and regional office still seems to take place in this way! It is hoped that the PCs in branches will act as platforms that will allow for the introduction of new customer services. This will include over the counter share dealing and foreign currency purchase.

- Corporate intranet. In the first instance, the Data Bank board see the introduction of a corporate intranet as a publication tool. By publishing the Data Bank staff directory on an intranet, many thousands of pounds per annum will be saved in production, printing and distribution costs of paper directories—the electronic version can also hope to be much more up to date! Other information to be published will be product information, training material, company rules and procedures, Health and Safety guidelines, in fact as much of the old paperwork as possible! Costs will be reduced, as employees will no longer have to receive and file this paperwork, and will be able to access the latest information much more quickly, using intranet searching tools. Once the company has established the publishing side of the intranet, they will have a valuable platform for the rapid introduction of new applications, using the standard Web browser as the user interface. Applications may be rapidly developed using state-of-the-art tools such as Java.

- The LAN interconnect/Intranet solution will be modelled on the successful ManageNet service. Data Bank realise that large scale networks are at their most efficient if the number of different protocols they support is minimised. Hence they will migrate to Novell 5, which can use TCP/IP for networking, instead of IPX.

- Legacy support and cost reduction. New systems must be capable of supporting ATM devices and their associated cluster controllers. The old (and increasingly unreliable) cashier terminals will be replaced with LAN connected PCs, able to carry out both banking transactions and office automation functions. The bank would like to eliminate support costs associated with running its own TDM multiplexer network, and also eliminate the costs associated with support, maintenance and software licences of the regional FEPs.

- Regional office rationalisation. Eight of the regional offices will close, with support functions being consolidated at the remaining sites. This will, of course, result in the loss of accommodation for some of the regional FEPs. The remaining regional offices will provide an increased level of back office processing for bank branches, so branches will need to communicate with their regional office as well as the host site. They are also likely to be relocated from their current city centre locations to cheaper accommodation in the outskirts of the City. At this stage, Data Bank has made no firm decisions as to which regional offices will remain, and where the remaining ones will relocate to.

- Travel reduction. Data Bank spends a great deal of money on travel, as its managers and regional office staff frequently meet at the headquarters office. To reduce costs and improve their environmental credentials, Data Bank are keen to fully exploit video conferencing. They plan to set up studios at the HQ and each regional office for meetings and training events.

- Call centre. At present, Data Bank customers would call their local branch if they have a phone query. This leads to variable levels of service, depending on availability of branch staff. Data Bank can improve customer service by directing all calls to a single national number, and have calls handled by a central call centre. Fewer staff will be needed at the branches, and it will be easier to plan the number of staff needed at any one time, to give consistent service at the call centre. The call centre can forward the more complex queries to the appropriate regional office or branch as necessary. The call centre is strategically very important, as it will allow Data Bank to offer a range of modern phone-based services for banking, investments and insurance.

- Outsourcing the network. The bank's IT department will not be allowed to grow, and needs to become more focused on developing and delivering bank applications, rather than running the communications network. A significant degree of outsourcing will be inevitable. They will be looking for the network provider to present an optimised service, using the best mix of networking technologies to provide high performance networking at least cost. Because the plans include removal of the existing TDM network, the outsourcing supplier must offer a solution for both the voice and data needs of Data Bank.

- Other cost reductions. Reduce costs and enhance reliability of the international dealing network, by migrating to a managed data network. Also sending external calls over IP/telephony server to reduce call charges.

- Electronic banking service. This is a major new initiative to allow Data Bank's business customers on-line access to their accounts. Access to this service will be over the public Internet. Data Bank will establish a web based front end to their banking systems. Users will access this service using their own inter-connected PCs. To help potential users who do not have Internet accounts, Data Bank is considering becoming a 'Virtual ISP.' Their outsourcing partner will provide the standard Internet Service Provider offering, rebadged as though it were a Data Bank product.

- Future proof. Networking technology is evolving rapidly, as are the types of applications that Data Bank may wish to run over their network. To ensure that Data Bank can maintain a competitive edge from their new network, they are seeking assurances from the supplier that the design is future proof. Also, to satisfy their shareholders, Data Bank want to be seen to be using state-of-the-art technology, including the latest Asynchronous Transfer Mode network services (ATM—not to be confused with the Automated Teller Machines!).

10.4   DATA BANK—THE PROPOSED FUTURE NETWORK

Having described Data Bank's existing network and future requirements, we can now go on to look at some actual solutions for their requirements. It is important to realise that there will always be more than one way to solve such a network requirement. Many factors will lead to the choice of the final solution, not least of which will be the availability of the different network technologies from possible suppliers, and the price that will be charged for each.

Given that there is no single correct solution for the bank's requirements, we have tried in this section to present solutions using a range of contemporary technologies, all based on actual networks of which the authors have personal experience.

10.5   INTERNATIONAL DEALING NETWORK

Let us first examine a solution for the international dealing network, as this is perhaps the simplest of the Data Bank requirements. It is similar in many respects to the case study presented in Chapter 2 of this book, and many of the same arguments apply in the design.

Architectural considerations

The following architectural aspects are of key importance to this design:

- DIY v. outsource. The existing network is very much a DIY affair. Data Bank wish to outsource the network and the operation and management of the associated routers. As part of the deal, Data Bank will be looking for the network supplier to 'buy and lease back' the existing routers (this will only be possible if the chosen network supplier has the experience and systems required to manage the existing type of router. The alternative will be for the supplier to replace the existing routers with models they are able to support, but this, of course, will increase the price of their bid).

- Solution technology platform. The existing solution is based on long haul leased international leased lines, but Data Bank is finding these too expensive, and not sufficiently reliable. X.25 networking was considered, and while this has good presence internationally, it was thought unable to offer the bandwidth required. Also, X.25 networks are normally implemented using store and forward techniques over modest trunk bandwidths, which increases latency (cross network delay). High latency solutions are not well suited for transaction-oriented client server applications as will be used here. We must also keep in mind that the current solution, based on 64 kbps leased lines will be providing ample bandwidth and minimal latency—the customer is unlikely to accept a replacement solution with a significantly lower performance!

Data Bank therefore decided to use Frame Relay, which can provide the required bandwidth, low latency and it has wide international availability. (At this time, availability internationally is nowhere near as widespread as X.25 networks, but several multinational network supply companies were able to provide service in all countries of immediate interest to Data Bank.)

Physical design

Site types

This relatively simple network has two site types: the central hub site in London and branch sites in other countries.

Traffic analysis

In this network, we know that all traffic flows are between the remote branch sites and the London hub site. There is no direct branch to branch traffic requirement. Traffic consists of a number of different TCP/IP protocols. What WAN bandwidth do we need? The existing network uses 64 kbps links to the remote offices. This bandwidth has proved more than satisfactory for all existing applications. The customer's engineering staff have captured data from the London router of the existing network. This shows that peak traffic from any one site does not exceed 12 kbps (the router averages the level of traffic on a link over a five minute period, and a graph was prepared from these data points during the course of a typical trading day). The customer has also made a number of other checks on router performance—particularly amounts of free buffer memory and router processor utilisation—this showed the router to have ample free memory and the CPU never exceeded 10% utilisation. Using Frame Relay will result in a small increase in CPU loading due to the link management protocol (LMI) that is used with this technology. Since the hub router is currently lightly loaded, this will not be a problem.

As a result of these investigations, it was decided that 64 kps Frame Relay service at the remote offices (the lowest speed Frame Relay service generally available from suppliers) was more than adequate. The exception to this is the New York site, where 56 kbps service will be used (USA networks use a different hierarchy of digital data rates in their public networks to that used in Europe). These sites will need Permanent Virtual Circuits (PVCs) configured across the network to the UK hub site. The price for PVCs will depend on the Committed Information Rate subscribed. (CIR is the amount of bandwidth that the network is expected to deliver across a PVC under normal network operational conditions.) To keep costs to a minimum, 16 kbps CIR will be used for all PVCs, which should satisfactorily support the known existing traffic.

In the UK, it was decided to use 128 kbps for connecting the hub site. This is more than the sum of CIR, so why was this rate chosen? First, we have to remember that at least two additional sites are planned, so using 128 kbps offers plenty of scope for growth, without the need to re-provision the London site early in the life of the contract. Circuits of 128 kbps are usually provided using a fractional E1 (2 Mbps) service, and can quickly be reconfigured to provide higher speed should the need arise. Circuits of 64 kbps would not be upgradable in this way, and would need complete replacement. The second reason is that international Frame Relay connections are invariably slower than the equivalent bandwidth when supplied by a terrestrial digital leased line (it would not be unusual for a Frame Relay connection to offer a faster delay than an equivalent satellite private circuit, due to the significant extra delay introduced by the satellite path). This is because the leased line is a bit pipe, with bits being delivered at the remote end with only minimal equipment delay, plus propagation delay. On the other hand, Frame Relay uses packet based technology, which has store and forward delay. As a minimum, frames have to be fully received at the destination Frame Relay node to have their frame checksum validated, before the frame can be passed to its destination. Frame Relay is also a statistical multiplexing network, so a user's packet can experience queuing delay, while waiting for other user's packets to be sent over trunk links. By using a higher speed link at the London site, we can help to minimise the extra delay that will be seen when the user migrates to Frame Relay.

Network providers will allow CIR oversubscription at hub sites (that is, the sum of CIRs for the PVCs terminated at that site exceeds the access circuit bandwidth). While this can be OK if there is a mix of LAN type applications (file transfer, e-mail etc.), it is unwise to exploit this facility where traffic is predominantly transaction oriented. That is because oversubscription can lead to significant queuing delays at the Frame Relay node (the access circuit becomes a bottleneck) and hence poor latency for transaction-based applications. The initial, trivial design for the network is shown in Figure 10.6.

The design is fine, and will work. It also represents a least cost solution. However, the design fails to fully tackle the customer's requirement for enhanced network reliability. The Frame Relay network itself will provide some additional resilience over an international private circuit, as each node will have multiple diversely routed connections to other nodes (i.e. the Frame Relay network is implemented as a partial mesh network). If

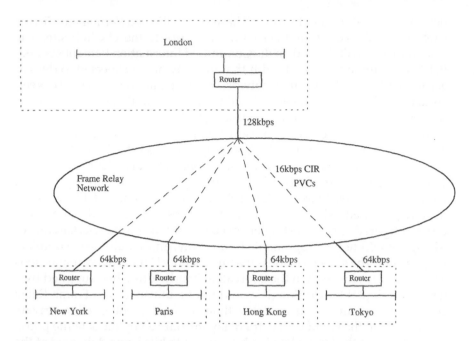

Figure 10.6 Initial Design for the International Dealing Network

any single trunk link fails in the network, the nodes are able to rebuild the customer's PVC via an alternative route. The network design in Figure 10.6 still has single points of failure in it, specifically the access circuits from the customer's sites to Frame Relay nodes are a key point of weakness and the single router in London, failure of which would impact all remote user sites.

To quantify the availability of the initial design, we need to find out availability figures for individual solution components. The proposed network supplier has kept statistics on component failure rates and repair times, and knows that the availability for a managed router provided by them is 99.9%. The access circuit availability varies from country to country, but the average is 99.5%. Once traffic reaches the Frame Relay core, it is passed over highly reliable switching equipment, located in the network operator's premises, where any faults can be quickly remedied. The network nodes are interconnected as a partial mesh for resilience. The network supplier quotes end-to-end availability of 99.95% for the network. DataBank need to know the end-to-end availability (i.e. what percentage of the time can a branch expect to be able to communicate successfully with the host site? This is given by the availability calculation illustrated in Figure 10.7.

End-to-end availability = .999 * .995 * .9995 * .995 * .999 = .987 or 98.7%

This is equivalent to 4.5 days of unavailability per remote site per year,

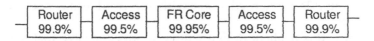

Figure 10.7 How we Look at an Availability Calculation

and this level is unacceptable to Data Bank. The design can be improved by adding a second router at London, with its own 128 kbps access path to a separate Frame Relay node (the second path should ideally be diversely routed, i.e. leave the building via a separate duct route). This router and its access circuit will be used as a standby, in case the primary path fails. Remote user sites will have their access circuits to the local Frame Relay backed up with an ISDN dial back up service (this may not be immediately available in all countries, as ISDN is by no means a universally available service yet on a world-wide basis). In order for remote user sites to connect via the secondary London router, they will need PVCs built to this router's Frame Relay access port. Again, 16 kbps CIR will be used. There will be additional PVC costs associated with this. Note that some suppliers recognise this need for resilient PVCs, and will supply the back up PVC at a lower price. These are often referred to as shadow PVCs. These require that the user makes an undertaking that the PVC will not be used for traffic under normal circumstances (obviously, if the network supplier also provides and configures the router, ensuring that this takes place will be their

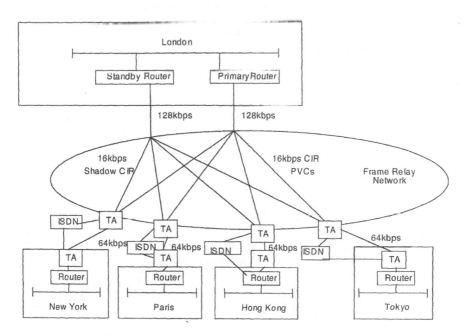

Figure 10.8 Resilient Design for the International Dealing Network

responsibility!) The enhanced design with improved resilience is shown in Figure 10.8.

Naturally, this resilient design will be more expensive that the initial design. The supplier would want to justify the additional costs to the customer. As before, we can use an availability calculation to show how the new design might be expected to improve end-to-end availability. The design introduces some new components. The ISDN switching Terminal Adapter, which is used to switch from the fixed access link to a switched ISDN back up call when the fixed circuit fails, has an availability of 99.95%. The ISDN services in the various countries have an availability of 98% (with such a service, not only does the infrastructure have to be working, but the back up call has to succeed when required. If the call is made at peak times of day, it may fail due to network congestion!). The availability chain is shown in Figure 10.9.

In the more complex availability calculations such as this, calculate the resilient sections (where two or more components are in parallel) first.

For the ISDN and access circuit, we have

$$0.98 \parallel 0.995 = 1 - ((1 - 0.98)*(1 - 0.995)) = 0.9999$$

For the two access circuits and routers at the host end, availability =

$$1 - ((1 - 0.995*0.999)*(1 - 0.995*0.999)) = 0.999964$$

End-to-end availability is now given by the series product

$$0.999 * 0.9995 * 0.9999 * 0.9995 * 0.9995 * 0.999964 = 0.9974 \text{ or } 99.74\%$$

This is equivalent to just 1 day outage per year, which Data Bank are prepared to accept.

Selecting the routers

The branch sites will have a single Ethernet interface and a single serial interface for the Frame Relay connection. The throughput is low for a router, so a simple low cost branch router is required. It is most likely that

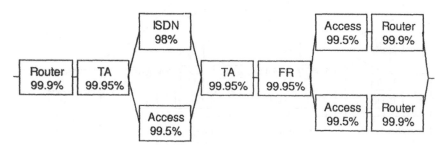

Figure 10.9 Another Look at Availability

the existing branch routers used for the private circuit solution could be taken over and reconfigured for Frame Relay operation.

Similar arguments apply at the London hub site—except that, if the resilient solution is adopted, the network supplier will have to provide an additional router.

To give a 'real world' example, the Cisco 2501 router would be suitable in all cases above.

Logical design

Now we have settled on a physical network design for this customer, we can think about the logical aspects, i.e. how the network works as a whole system.

Addressing

The only protocol being run in this network is the TCP/IP suite and so the key addresses are IP addresses. These are 32-bit addresses (usually written as four decimal numbers separated by dots, each number representing 8 bits of the address). Because of its association with the Internet, the IP address space is centrally controlled, and users would normally apply for a registered address space to the NIC organisation in the USA or RIPE in Europe.

While this customer wishes to outsource management of the network, they still wish to retain control of network address administration (after all, these addresses have to be installed in many different end systems and it would be inconvenient to agree every end system change with the network supplier). The existing network has a registered 'class B' address. This means that the most significant 16 bits will be a fixed 'network' address. Let us say it is 145.123.x.y. The user is now able to allocate the least significant 16 bits to the end systems or hosts. In routed networks, the user will further subdivide the host address space by creating subnetworks. These are individual LAN segments or wide area network segments that are interconnected by the routers. The user chooses how many bits of the host address space to use to represent the subnetwork. In this case, the customer decided on 8 bits (routers are told about this division by associating addressing information with a 'subnet mask'—in this case 255.255.255.0—which says only the least significant 8 bits represent end systems).

The customer has to assign subnetworks to each site LAN and also one for the Wide area network (145.123.1.x is used for the WAN in this example). The Network supplier will also require the customer to allocate a host address to be used on each router port as shown in Table 10.1.

Table 10.1

Site	Subnetwork	Router LAN port	Router WAN port
London Main	145.123.2.x	145.123.2.254	145.123.1.1
London Standby	145.123.2.x	145.123.2.254	145.123.1.2
New York	145.123.3.x	145.123.3.254	145.123.1.3
Paris	145.123.4.x	145.123.4.254	145.123.1.4
Hong Kong	145.123.5.x	145.123.5.254	145.123.1.5
Tokyo	145.123.6.x	145.123.6.254	145.123.1.6

Routing

The first step in setting up routing will be to load the routers with information about the Frame Relay network connections. The Frame Relay port will need to be configured with its IP address, and also told of the PVCs available on the connection (each PVC has an identifier called a Data Link Channel Identifier or DLCI—the network supplier allocates this address to each PVC). The router will need to determine which IP address can be reached at the other end of a PVC. This can be configured manually, or some routers are able to discover this automatically by performing a Frame Relay 'inverse ARP' procedure (a request is sent to the remote router on that PVC and the remote router responds with its IP address).

A routing protocol now needs to be set up to allow the LANs to interconnect. First, the IP addresses of the LAN ports must be loaded into the router. After this has been done, the routing protocol can be configured. This will then proceed to automatically build the required routing tables in each router.

In a small network such as this, the simple RIP protocol is quite sufficient. The routers would be set up to allow the routing updates to pass out into the LANs as well as over the Frame Relay network PVCs. This is because the Unix servers are able to receive these routing updates, and use them to determine the address of an appropriate router to use when passing traffic to a remote destination. This is particularly important in the case of London, where the server may need to switch routers if the primary path or its associated router fails.

We said that one of the routers and its associated PVCs would be a standby, and not normally carry user traffic—how is this arranged? This can done by configuring the routing protocol in the host site standby router to add a hop count offset to any routing updates sent and received over the secondary path. This would make destinations via this route appear to be more hops away than the primary route, and hence that path would not be selected by the routers (unless of course the primary path fails). This is illustrated in Figure 10.10.

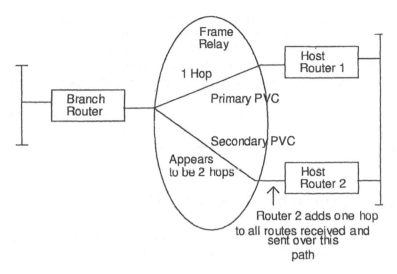

Figure 10.10 On the Hop

Management design

All routers are to be managed centrally by the network operator. This is achieved from a network management centre which is attached to the Frame Relay network. The management system will have a PVC built to each router to be managed.

Management will be via the SNMP protocol, which runs over the IP protocol. This means that the network provider will have to provide IP management addresses to each router (from the address range recognised by the management system routers), in addition to the customer's addressing scheme. The router is able to have two addresses associated with the Frame Relay port in this way, and one of them is declared as the primary address and one the secondary address. If a router ever needs to start from a position where it needs to download code or configuration from a management server, it will always tend to use the primary address for this purpose (exact method of operation will obviously depend on router manufacturer!). Therefore, the management address is designated primary, and the customer network address, secondary.

Futures

The main concern for the future with this network is scalability. This is in terms of the number of sites and the bandwidth needed at the host site.

Extra branch sites can be readily added, simply by cloning the design and configuration of existing branches. Once connected to the Frame Relay cloud, the new branch will need its two PVCs, the primary and secondary, built in to the host site. The host site will be seen to be told of the new PVCs, and once this is done, the routing algorithm will 'discover' the new site, and traffic can start to flow.

Additional bandwidth can readily be provided at the host site by asking the network supplier to upgrade the speed of the fractional E1 service (i.e. by using more of the 30 64 kbps time slots available on the E1 circuit).

10.6   THE DATA BANK BRANCH NETWORK

Given that the proposed Data Bank branch network is a large and complex, it is especially important that we systematically work through our design process as follows:

- analyse the site types;

- analyse the traffic flows between sites;

- decide basic network architecture and technology;

- determine if design is sufficiently reliable. If not, enhance design to increase resilience;

- dimension network;

- determine router requirements at each site type;

- carry out logical design for network and determine if this has any impact on the physical design.

Summary of site types

All the Data Bank site types were introduced in section 10.2. To recap, the site types we have to interconnect are:

- Host sites—there are two of these.

- Regional offices—there are 16 of these. We know from section 10.3 that eight of these are to close. It will be important to press Data Bank for the earliest possible decision as to which sites are to remain and which are to be closed. The two cases will need to be treated differently the design process, with a focus on short-term solution for the sites to be closed and more on a long-term evolution for those that will remain.

- Branches—there are 400 of these, of which 100 are in large towns and therefore handle significantly more business.

- Call centre—closer examination of the requirements show that the bank intends to establish this. It will be located in an area of the UK where labour costs are low. In order to offer customer service, this site will need connection to the Data Bank host sites.

Analysis of traffic flows

When we examine the requirements given in section 10.2, we notice is that there is very little quantitative information, particularly concerning application traffic flows. Without this information, it would be very difficult to dimension the new network correctly. The first task of the network designer will therefore be to discuss further with the customer the quantitative aspects of the requirements. This is not always easy for the customer, especially when many of the new applications to be run on the network will still be at the drawing board stage. The main thing for the network supplier is to get from the customer a signed off set of assumptions concerning traffic levels. If subsequently the assumptions are found to be too low, the network supplier will be in a good position to ask for more money when supplying extra bandwidth! In further discussions with Data Bank, the following has been discovered:

- The flow of SNA traffic between branch and host is well understood. At peak times, a busy city branch (there are 100 of these) completes a transaction every 10 seconds (this represents the typical large branch— small branches average half this traffic load). A transaction consists of two message pairs, typically of 80 octets to the host and 250 back. This gives an average throughput of 400 bps in the host to branch direction. It is known that, as the branch terminals are replaced by PCs, the transaction sizes will increase considerably, as more information is downloaded to be displayed during each transaction. It is estimated that transaction traffic will increase by a factor of 4 to 1600 bps. The typical smaller branches (300 of these) have half the amount of traffic as the busy city branch.

- At the host, the 400 branches therefore generate a worst case host to branch flow of 100 kbps, this will increase to 400 kbps (1600 bps*100 + 800 bps*300) with the new transactions.

- The host carries out a daily batch transfer from each branch as part of the daily reconciliation process. The file size averages 1 Mbyte for small branches and 2 Mbyte for larger branches. There is a critical four-hour window during which this must be collected, and the bank prefers to complete it in 90 minutes, to allow for any recovery. This requires a branch throughput of 1560 bps at small branches and 3120 bps at large branches. This is a total bandwidth of 780 kbps at the host. This transfer is not expected to increase significantly with the new applications. It can be seen that this batch transfer is more demanding than the transaction traffic (many networks are found to have more demanding file transfer throughput requirements than transaction bandwidth).

- Host site to host site flow peaks at around 1 Mbps.

- Regional offices do a great deal of back office work, and this is batched up for an overnight transfer to the host site. The regional offices also provide print shop facilities for statements and other customer mailings. The host also sends this information in batch overnight. The transfers here are allowed three hours to complete and have a worst case of 50 Mbyte from regional office to host and 20 Mbyte from host to regional office. Transfer only occurs in one direction at a time. This needs some 55 kbps of bandwidth capability in each direction, and with the planned eight regional offices, 352 kbps at the host.

- For the OA requirement, there will be e-mail post offices at the regional and host sites. Any traffic between regions will flow via the host site post offices. Mail will flow between regional and host post offices during the working day. The flow is not expected to exceed 100 kbps at the host at any time with each regional post office.

- Branch to regional office will become considerable, consisting of e-mail flows during the day and various overnight batch transfers. E-mail is expected to be as high as 20 kbps at a branch when the manager logs on to read mails (irrespective of branch size). The most demanding download from regional offices is expected to be major branch software upgrades. This is expected to be around 100 Mbyte in size and would be rolled out to all branches over an eight-hour period on a Sunday night (when other regular batch transfers do not occur). This needs a throughput of 29 kbps at a branch, and 1.74 Mbps at a regional office supporting a worst case of 60 branches.

- For the call centre, it is estimated that 500 kbps of bandwidth is needed to the host site by day and 250 kbps out of normal banking hours.

All figures above reflect the average traffic load expected during the peak hour for the relevant daytime or night time traffic flow.

The traffic flows above can be summarised in a traffic flow diagram (Figure 10.11). This shows peak traffic rates (in the worst case direction) for branches and regional offices, together with total bandwidth needs at the different site types. This is given separately for day and night traffic. Note that in the figure, all traffic directed to the live host should be able to

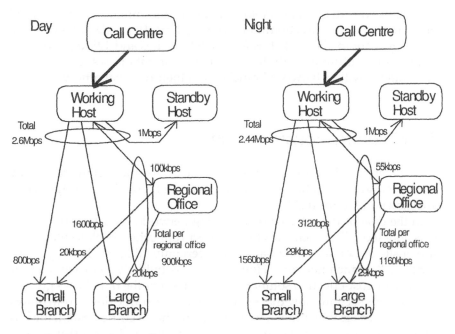

Figure 10.11 Night and Day

be automatically switched to the standby host if the live host fails or is taken down for routine maintenance.

As Data Bank wish to cease use of their old TDM network, an alternative method must be found for carrying the voice traffic. The data network must provide bandwidth for this voice traffic. Data Bank have the following requirements:

- Each regional office will have a PABX. At peak times, 30 voice channels are sufficient for traffic within the organisation.

- There will be two main switches, located at the host sites.

- The call centre will handle most calls locally, but up to 60 calls may be forwarded to regional offices or branches at any one time.

Figure 10.11 summarises the known, quantifiable data flows. Data Bank are keen, however, to have a future-proof network. They need branches to have spare capacity on their branch links for new applications, an example of which might be interactive Web-based customer terminals (multimedia customer kiosks) which can be used to provide information to customers about Data Bank products. Data Bank are also excited about the prospect of reducing branch telephony costs by moving some of the traffic onto a voice over IP service. This would take branch traffic to the regional office PABX.

Data Bank consider that technology in this area is immature, and they will only proceed in this direction after careful technical trials. Data Bank decide that they want to allow 64 kbps for small branches and 128 kbps for large branches for future growth. It is assumed that data flows for these new applications will be between branches and their regional offices.

Architectural considerations

Two widely used solutions can be considered for the architecture of the Data Bank solution: a host centric design and a hierarchical design. These are shown in Figure 10.12.

In the case of the international dealing network, a host centric design was adopted, all branch sites had a direct connection back to the central server site in London. Data Bank could certainly adopt such an approach for the UK branch network. However, this is unlikely to prove the most efficient or cost-effective solution. Examination of the data flows given in the previous section show that significantly more traffic flows from branch to regional office than branch to host site. In a host centric design, the host sites would need very large WAN connections and powerful routers,

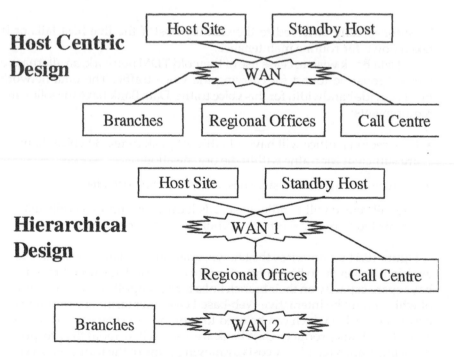

Figure 10.12 The Two Options for Data Bank

simply to act as a forwarding site between branch and regional office!

The alternative approach, which Data Bank select to use, is the hierarchical design. It would make sense to have branches clustered around their regional offices, with routers at the regional offices concentrating and forwarding traffic to the host site. This design approach has a number of additional advantages:

- It will facilitate the logical design of the network—such large networks have to be divided into routing areas, and this can be conveniently done on a regional basis.

- There is the opportunity to select different WAN types for the regional clusters and the network core.

The hierarchical design is not without its drawbacks—it is very sensitive to any plans to close or relocate regional offices. Data Bank will be forced to make an early decision on its regional office rationalisation programme, and risk network disruption if any of these regional offices are closed or relocated in future. In some cases, it is desirable to ask the network supplier to host regional concentrator routers at their network node sites, and link the regional office into these concentrators. This permits a regional office to move with much less impact on its branches.

Selecting the WAN technology

Choice of WAN technology is influenced by many factors, already described in Chapters 5. Data Bank would carry out 'thumbnail' designs using different technologies and have these costed.

For the core WAN, a range of technologies are typically used by different banks. These include TDM/digital private circuit solutions, Frame Relay, SMDS and ATM. Data Bank need to combine both data and voice circuits into their core WAN. ATM is the technology best able to support this. It can offer virtual circuits designed for voice (constant bit rate) and virtual circuits designed for bursty data (variable bit rate) over a common access circuit. ATM is a high bandwidth service and Data Bank will need to buy 34 Mbps access circuits at each site (the lowest speed that their proposed service provider offers). This may seem excessive to them, but it will give a very future proof design, and much of the cost for the service will be based on the committed bandwidth of the different virtual circuits, so Data Bank need only buy what bandwidth they need for these. ATM multiplexing devices will be needed at each regional office and host site, to combine the voice and data streams. The ATM multiplexors can also be used to provide high quality paths over the ATM core, which will be required to support the proposed video conferencing studios.

For the regional concentrator networks, banks would typically use either Frame Relay or digital private circuits. Data Bank have done a costing exercise with potential network suppliers and have found the digital private circuits to be the most reasonably priced solution. (For relatively short distance national connections, this is often the case, as Frame Relay services have to be delivered over a digital private circuit and always include the price for this in the overall service charge.) Using digital private circuits between the branch and the regional office will:

- Guarantee bandwidth for the branch. There is no statistical sharing of bandwidth there would have been if Frame Relay was used.

- Minimise network delay. Frame Relay uses frame based store and forward switching which adds to network delay. Digital private circuits add delays of only a few bits. Low network latency is important as Data Banks main business transactions are carried out interactively between branch equipment and their host centre mainframe.

In selecting the type of digital private circuits to be used, Data Bank specify the following:

- They should be delivered over optical fibre bearers whereever possible. Optical fibre delivery tends to be more reliable than traditional copper-based circuits, and the enormous bandwidth potential of fibre will provide future proofing for the design.

- A fractional E1 service should be used. This offers the customer bandwidth in multiples of 64 kbps time slots (Fractional T1 in the USA, in multiples of 56 kbps). This allows the customer to upgrade branch bandwidth relatively simply, if needed, without the need to re-provision the access circuit.

- Aggregated delivery at the regional office—this allows multiple branch circuits to be presented on a single 2 Mbps E1 bearer (up to a maximum of 30 slots of 64 kbps). This greatly reduces the amount of equipment and the complexity of the router and its wiring at the regional office.

Resilience considerations

Another key aspect of requirements that needs to be amplified is that for network availability. Clearly, the network is absolutely essential for Data Bank operations, and if the network is to be outsourced, they will insist on a challenging Service Level Agreement with the supplier. The following requirements have been given:

- Branch availability (measured over the full population of branches, on a monthly basis) is not to fall below 99.9%. The supplier will have to pay rebates if service falls below this level.

* 80% of all branch faults must be fixed within four hours. Any site out of service for more than one working day in any month will have its site rental rebated for that month.

* The network must be designed so that host availability is better than 99.95%.

* Regional office availability must be 99.9%.

Given that access circuits for branches, regional offices and host sites seldom have availability of better than 99.9%, the end-to-end availability of the service will be much less than this (the availability product would include access circuits at either end of the link and routers). It is clear that resilience must be added to the design.

For branches, Data Bank decide to use ISDN. For the small, relatively low traffic sites of a network, duplicated access circuits over diverse routes are a luxury that cannot be afforded.

All other sites in the Data Bank will get resilience through the use of dual WAN links, offered over diverse routers to different network node sites.

Summary of architectural design

Before proceeding to look at the physical design of each site type, let us summarise the overall architectural design. This is shown in Figure 10.13.

Completing the physical design and dimensioning connections

Branches

The first decision Data Bank made was how the branches should be wired for LANs. As branches are relatively small, Data Bank intends to use a single 10 Mbps Ethernet LAN. They choose to 'flood wire' each branch using Unshielded Twisted Pair (UTP) cabling. Each work position is provided with a dual socket and two UTP cables leading back to the comms room (one is a spare in case a cable fails). If Data Bank moves toward use of voice over IP, the second circuit could be used for IP telephones. PCs with Ethernet adapters can be plugged in to these sockets as and when required. The UTP cable selected is certified for 100 Mbps operation, should Data Bank upgrade to using the higher speed form of Ethernet in the future.

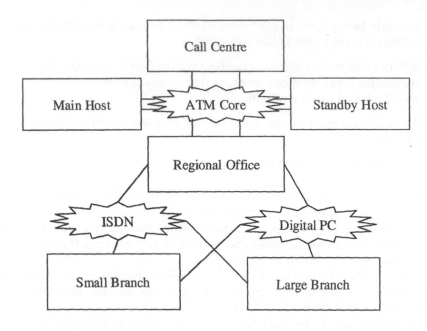

Figure 10.13 The Big Picture—Overall Architecture

In the comms room, UTP cables are connected to the ports of an Ethernet hub. At this stage, Data Bank elect to use a low cost non switched hub, this could be replaced in future with a 10/100 Mbps switched hub if branch traffic grows sufficiently to justify this.

The comms room houses the local server systems that Data Bank will use at branch level. There is a Novell server for local office automation, and also an OS/2 Communications Manager, which provides concentration of SNA sessions, acting rather like a cluster controller. The comms room is also the home of the cluster controller device that concentrates traffic from ATMs and cashier terminals.

When planning the WAN connection from the site, we need to consider the information given in an earlier section. This shows that the peak known data rate for Data Bank branches occurs outside branch trading hours—30.56 kbps for small branches and 32.12 kbps at large branches. It is unwise to permit peak traffic loads to represent more than 70% utilisation of available bandwidth. Much more than this results in significant queuing at the network interface, leading to unpredictable network delays and the risk of data loss. (Remember that in IP networks it is perfectly normal to discard data packets if maximum queue lengths are exceeded—this is used to trigger the layer 4 TCP congestion control mechanisms, which will back off and offer less traffic to the network until the congestion is cleared.) Allowing for this 70% 'rule', we therefore need 44 kbps at small sites and 46 k at large sites. Bandwidth for digital private circuits is normally provided in multiples of 64 kbps (56 kbps in the USA). Therefore a 64 kbps path is the minimum suitable bandwidth for each site type to allow for today's predicted traffic. However, Data Bank has made it clear that it needs an extra 64 kbps at small branches and 128 kbps at large branches to allow for future expansion. This means that the network supplier needs to provide 128 kbps at small branches and 192 kbps at the larger branches.

Data Bank were not satisfied with the availability quoted for the private circuit, and have asked the network provider to provide ISDN dial back up. To keep costs to a minimum, a single B channel only will be used, providing 64 kbps. This is sufficient to provide basic branch facilities, but some of the more advanced services, like the multimedia customer kiosks, may have to be withdrawn during outage of the primary network connection.

To complete the solution for the branches, we need to select an appropriate router to interconnect the LAN and WAN. The branch router has to support:

- An Ethernet connection.

- A serial X.21 interface for the digital private circuit.

- A serial V.24 port for the SDLC cluster controller.

- An ISDN Basic Rate Interface.

An appropriate real life model to use here is the Cisco 2503. This has been specified with maximum RAM memory and FLASH ROM program memory to future proof it against future applications and software upgrades. Top of the range Enterprise software is specified—this is needed to support IP and IPX protocols used, together with the legacy SDLC interface. The overall branch design is summarised in Figure 10.14.

Regional office design

Let us start with traffic arriving from the branches. We know that there are 400 branches to be served by eight regional offices. It is unlikely that the regions will be set up so that there is an exact division of branches between regions. We discover that the worst case situation is 60 branches per region; of these, 25% will be large branches (needing three 64 kbps time slots) and 75% are small (needing two time slots). That gives a total requirement of 135 time slots. Branch traffic will be delivered to the host on E1 digital circuits, supporting 30 channels each. The minimum number we need is five.

In order to make a resilient design, however, branches will be split across two routers—30 branches per router (worst case). Supplying three

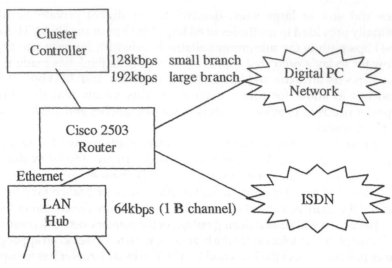

To branch PCs and local server

Figure 10.14 The Branch Design

E1s per routers will give a symmetric design with spare slots to allow for growth. If the number of branches per regional office is less than 50, then four E1 circuits will provide sufficient capacity.

ISDN back up is required. If a site is on ISDN back up, it will only use a single B channel giving 64 kbps. One ISDN Primary Rate service, supporting 30 B channels per router gives capacity for full branch ISDN back up. The service is designed so that a branch has its primary data circuit terminated on one router, then its ISDN back up is to the alternative router.

Regional offices use Token Ring technology. Each of the two regional office routers connect to two backbone rings. (There is also a second ring, used exclusively by the routers for switching traffic between each other.) The backbone rings run between building floors. On each floor there is a Token Ring bridge and MAU associated with each backbone. The MAUs are interconnected into a single ring. Users connected to the floor ring are able to have traffic delivered via either backbone, creating a resilient solution.

The regional office is connected to the ATM service at 34 Mbps. This is done using resilient SDH access circuits. Two ATM services to different serving nodes are provided for resilience.

This ATM service has to be used by the routers, regional office PABX traffic and video feeds from the proposed video conferencing facilities. An ATM service multiplexer (e.g. Cisco LS1010) is used to combine these traffic streams. Only a single unit is being deployed initially to keep costs down. It is a single point of failure, but the designer believes the unit has high reliability. Design trade-offs like this sometimes have to be made.

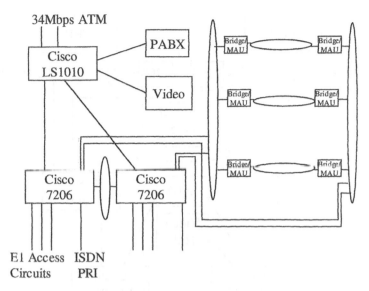

34Mbps ATM

Figure 10.15 The Regional Office Design

Figure 10.15 summaries the regional office design.

When selecting the routers for this design, we must look for devices that can support:

- Three E1 and 1 ISDN PRI ports.

- One ATM port.

- Three Token Ring ports.

- Substantial processing power, required to handle routing protocols and some DLSW processing to transport a certain amount of IBM SNA traffic that is generated on the regional office rings.

The Cisco 7206 router has been selected to meet these requirements.

Host centre design

Each host centre will have two ATM links for regional and branch office traffic. The sites are interconnected by a pair of point-to-point private circuit services, able to carry 155 Mbps ATM traffic. At these sites will be a hub PABX to link regional offices. The main links to the public telephone network also attach to these PABXs. A video conferencing hub is also provided at the host site. As with regional offices, the ATM traffic is switched using LS1010 devices—this time we have two for resilience.

Data traffic terminates at a pair of Cisco 7500 routers. These feed Token Rings housing key intranet servers. There are also rings to feed the IBM

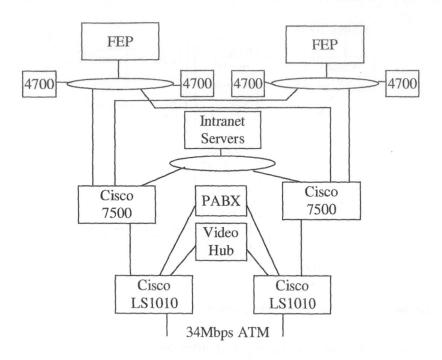

Figure 10.16 The Design for the Host Centre

mainframe FEPs. Substantial amounts of processing are required at host sites to handle the DLSW encapsulation for branch and regional office SNA traffic. This is handled by four Cisco 4700 routers attached to the FEP rings. This design is summarised in Figure 10.16.

Call centre design

The call centre site will have a design based heavily on that of a regional office, but with the following two differences:

- There will be no E1 or ISDN PRI connections, as the call centre does not act as a branch concentrator site.

- The PABX at this location will be much larger than that at regional offices, reflecting the high call volumes. The PABX will also have a large number of connections to the public phone network.

Logical design

WAN configuration

The key task here will be to design and provision appropriate ATM logical channels between the regional offices, call centres and host sites. The ATM network must carry two distinct types of traffic: for PABX voice interconnectivity and for data interconnectivity. These two types of traffic have different characteristics and are therefore best served by two separate logical channels with appropriate configuration.

The voice traffic requires guaranteed bandwidth, low latency and minimum jitter, to ensure good quality voice performance. This is achieved by using an ATM Circuit Emulation Service, which uses a Constant Bandwidth (CBR) logical channel and AAL1 encapsulation. The Data Bank PABXs handle voice as 64 kbps digital channels, so the ATM bandwidth subscription required between any two points will be based on the product of 64 kbps and the maximum number of simultaneous calls we plan to support over that route. In an earlier section it was stated that regional office PABX have 30 channels to the host PABXs, giving an ATM bandwidth of 2 Mbps (the PABX connection will actually have 32 channels—two extra channels being for signalling and timing. Note also that the actual ATM bandwidth subscribed will need to be higher than this to allow for the ATM cell overheads × 53/48 = 2262 kbps.

The data traffic is expected to be bursty in nature. Bandwidth guarantees, low latency and low jitter are much less of an issue. In this case we can use the ATM Variable Bandwidth (VBR) logical channels. Details of VBR services vary from one service provider to another, but are based on the concept of being able to subscribe to a minimum guaranteed bandwidth plus the ability to burst above this within defined parameters. Excess burst traffic is generally discarded by the network (though some VBR services may tag excess bursts for discard only when the network is congested). The ATM CPE at the customer's site is therefore required to shape traffic within agreed parameters to prevent loss in the network.

Subscription to a VBR logical channel requires the customer to select a guaranteed throughput, known as the SCR. This should be set at a sufficient level to cope with average peak hour traffic. To cope with statistical bursts above the mean, the user also subscribes to a peak rate (PCR) and a maximum burst size MBS for which this peak flow rate is permitted to be susta'ined. In a typical network, PCR would be set to twice the value of SCR and a burst size of up to 200 cells would be permitted. The SCR subscribed must, of course, be based on the application bandwidth required multiplied by the ATM cell overhead, 53/48. SCR and PCR is generally sold in units of 1 kbps.

A key advantage of using ATM is that we can configure our logical channels and only pay for the bandwidth required now. Provided the access circuits have the capacity, we can subscribe for more bandwidth later as new applications come on stream. Data Bank intend to fully exploit this capability and will only initially subscribe for sufficient bandwidth to carry existing applications. The bandwidth subscription will be increased in stages, as groups of branches migrate to the new PC based end systems and as new applications are phased in.

The logical channels required are:

- Host site to host site needs 1 Mbps, so 1105 kbps is subscribed.

- Call centre to each host site—500 kbps to cope with daytime peak flows, so 553 kbps SCR is needed.

- Regional office to each host site—this must hand regional office to host traffic, plus the concentrated branch traffic. Our worst case regional office has 60 branches, 15 large and 45 small. The night time traffic requirement (45*1560 + 15*3120) + 55 kbps = 172 kbps slightly exceeds the daytime requirement. Regional office to host SCR is thus 190 kbps.

Addressing scheme

The next key area for Data Bank will be to decide on an IP addressing scheme for the network. As this is a large private network, Data Bank have chosen to use the Internet unregistered class A address space, 10.0.0.0. As the company has not had an IP network to its branches before, it can create an addressing scheme from scratch, and ensure that it is designed for maximum routing efficiency.

As shown in Figure 10.17, the network will rely on a multi-area OSPF routed design. OSPF can carry out address summarisation, keeping its overhead to a minimum, but only if the addressing scheme is designed cleanly. To achieve this, Data Bank will reserve the second 8-bit field of the IP address to represent the OSPF Area. Thus addresses will have the format 10. < area number > .0.0. This allows us to have 254 areas.

A sensible maximum number of routers in an OSPF area is 50. We can use 6 bits of the third 8-bit field to represent up to 62 routers associated with the area. This leaves 10 bits in the addressing field for assignment to end devices within a router's subnetwork, i.e. 1022 devices. The subnet mask will be 255.255.252.0.

Routing

This is a relatively large network, which can benefit from using a structured multi-area OSPF scheme. Such a scheme will minimise the amount of bandwidth consumed by routing updates and will be capable of rapid convergence following any failure of a core network link. The OSPF design is shown in Figure 10.17.

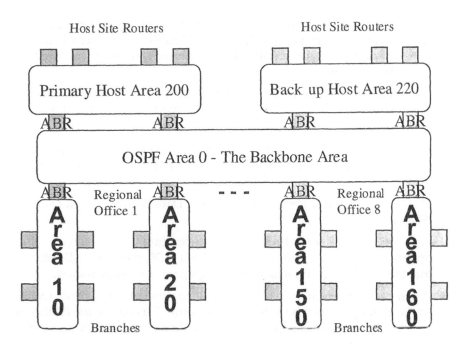

Figure 10.17 Routing Arrangements for Data Bank

Areas have been numbered in multiples of 10. This has been done to allow for any future growth in the network needing addition of new areas. Areas are interconnected at Area Boundary Routers (ABRs). In the case of the regions, the ABRs are the regional office routers and in the case of host sites, they are the ATM core access routers. It must be remembered that ABR duties require extra processing power and memory in a router, so high end devices should be selected for this role.

ISDN DBU

OSPF provides a good stable routing platform for the network as a whole. It is not, however, an appropriate routing protocol to operate over the ISDN back up link. Experience has shown that a more stable network can be achieved using a different routing protocol to manage the ISDN DBU—in this case Data Bank will use the RIP version 2 protocol. The network is configured as follows:

- Branch routers have a 'floating static route' to network 10.0.0.0 which directs traffic to the ISDN port. The floating static route has a higher 'cost' than the route over the primary access path provided by OSPF, so traffic will not normally flow over this route.

- Branch routers run a RIPv2 process which sends routing updates to a specific routing peer address located within the router which will be used to terminate the back up call. This router is normally reached via the primary circuit.

- When the primary route fails, customer traffic (or the RIPv2 update if there is no user traffic at the time) will have to use the floating static route instead.

- This traffic triggers the ISDN call to set up. The receiving concentrator will have to check the Calling Line Identifier (CLI) in the incoming message for security reasons. If all is well, a call is established.

- The branch can now send traffic into the Data Bank network over the ISDN line.

- For traffic in the reverse direction, the concentrator must receive a RIPv2 update. The resulting route to the branch via ISDN has to be redistributed into the OSPF protocol, which then propagates this information.

- When the primary link is restored, the routing protocols will restore traffic to this path as it is the lowest cost route. This means the ISDN link will see no traffic, and it is configured to time out and disconnect after becoming idle.

Protocol support

All new Data Bank LAN based applications use the TCP/IP protocol. The routers used in the Data Bank network are optimised for transporting this protocol. The key protocol support issue is for the IBM legacy protocols.

The IBM traffic terminates on FEP Token Rings, and will be transported using the LLC2 connection-oriented link layer protocol. This protocol has no network layer and has to be bridged across the network. Given that the core network is an IP routing environment, the best way to bridge the LLC2 traffic is to encapsulate it in TCP/IP sessions. This is achieved using the standards based DLSW protocol. LLC2 traffic is targeted to a link layer MAC address. When traffic arrives at a DLSW process, a communication takes place with peer DLSW processes. These send out local explorer packets to try to locate the MAC address. A TCP/IP session will be established with the peer that locates the MAC address, and all future LLC2 traffic for that MAC address will flow over that TCP/IP session. Because we have a resilient IP network underpinning DLSW and because TCP/IP is very tolerant of short link breaks and packet drops, Data Bank will have extremely reliable transport for the SNA legacy code.

Managing the TCP/IP sessions is CPU intensive, so networks of this type will have dedicated, high performance routers at the host sites, specifically to terminate DLSW traffic. Data Bank has a number of Cisco 4700 devices for this purpose. When configuring the network, each branch router will have to have configuration statements associating it with four host end peers, namely:

• The primary host working peer.
• A primary host back up peer in a second router (to cover the eventuality of the primary peering router failing).
• The standby host working peer.
• A standby host backs up peer in a second router at that site.

Both primary and back up DLSW peers can see the same host MAC address, and will report this back to the branch router. The branch router has to be programmed to give a higher precedence to its designated primary, and only use the secondary if it cannot establish a TCP/IP session with the primary.

One final point, IBM legacy traffic from the ATM cluster controller does not enter the branch traffic as LLC2, but as serial SDLC data. The router has to first convert the SDLC stream to LLC2, achieved by the Cisco SDLLC functionality. To make this work, the SDLC link will have to be assigned a pseudo MAC address, which needs to be unique for this network—yet another job for the network administrator to handle!

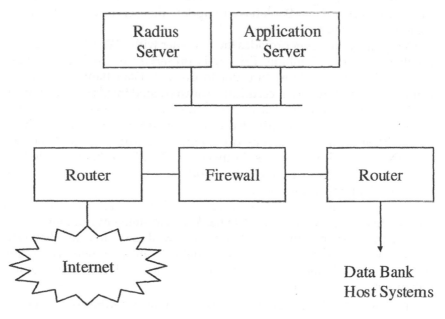

Figure 10.18   The Internet Banking Trial

10.7 INTERNET BANKING

Data Bank want to run a limited trial of Internet banking initially. They will offer a free of charge service to their 'Gold Card' customers. To run this trial, they will establish a non resilient server, located at the main host site. An outline design is shown in Figure 10.18.

The solution starts with a fixed (private circuit) connection to the public Internet. With a limited trial in mind, Data Bank have a low speed 256 kbps connection in the first instance. This is provided by an Internet ISP, which will also be able to obtain a registered class C address range for Data Bank to use. The ISP also provide and configure an Internet access router. Data Bank have registered an Internet domain name 'www.databank.com'. The ISP will configure the network so that this name is now recognised and translated to appropriate IP addresses by the Internet Domain Name Servers.

Traffic from the Internet is delivered to a firewall. Data Bank have elected to use Checkpoint's Firewall/1 solution, which is installed on a Sun Sparc server running the Solaris operating system. The firewall creates a DMZ (De-Militarized Zone). Traffic from Data Bank customers, of a type permitted by the firewall (e.g. HTTP traffic) is passed by the firewall into the DMZ. There is no ability for Data Bank customers to pass traffic directly to the Data Bank host systems.

In order to establish connectivity with the application server, customers have to log on using account number and password. This is handled by the radius server. Data Bank have to update this server each time a new customer is added to the service.

The application server provides the customer with a Web-based interface for their electronic banking. Privacy is preserved through use of SSL3 encrypted HTTP sessions. In order to do that, Data Bank will have to acquire a public key server certificate from a trusted third party Certificate Authority (e.g. Verisign). The application server is not just a simple Web server in this case. It is a dedicated piece of 'middleware' software that is able to integrate a customer based Web interface with Data Bank's legacy IBM SNA based host systems. In this case, Data Bank have selected the Edify product, which is integration software specifically aimed at the electronic banking application.

The firewall will be configured to specifically pass traffic from the Edify code on to a router within the Data Bank host system environment.

If the trial is successful—commercially as well as technically, then Data Bank will want to make the facility available to all customers. To do this, the network will have to increase in capacity and resilience. Design enhancements will include:

- Upgrade of ISP Internet connections—perhaps to 2 Mbps or beyond.

- Setting up a second server site, with traffic load balanced across the two.

10.8   VOICE OVER IP TRIAL

Data Bank is keen to examine the potential of voice over IP technology. It plans to start this with a limited scale trial at branch sites using the new network. Today, all telephony communication with the branch is via the public telephone network. Data Bank believe that costs can be reduced if some telephony traffic could, be transported instead over the new data network. Data Bank plan to use this approach exclusively for internal calls during the trial, and not for customer facing traffic. (Fewer customer calls are now being handled at branches anyway, as Data Bank increasingly direct incoming customer calls to the call centre.)

IP telephony systems digitise voice samples in the same way as conventional digital telephony. Instead of sending the samples as a continuous bit stream over a dedicated link or TDM timeslot, IP telephony sends sets of voice samples in small IP packets over the shared IP network.

When planning the trial, Data Bank will have to decide what telephony end systems to use at the branch. Four main options can be considered:

- Use the multimedia capabilities of Office PCs, and ITU H.323 compliant software such as Microsoft NetMeeting. This was rejected as most branch PCs do not have multimedia interfaces at this time. Use of the PC could be ideal for cases where a high degree of Computer Telephony Integration (CTI) is required—Data Bank would consider the use of this technology in its call centre.

- Use dedicated IP telephones. These are telephones with Ethernet interfaces, able to plug directly into branch LANs and deliver IP packets via the branch LAN.

- Use existing telephones and add an IP PBX. This acts as a small branch PBX, but instead of delivering calls to traditional trunk circuits, it delivers calls to the IP infrastructure. An IP PBX has to be supplied at regional offices (see below) so this technique will be used there.

- Use a router with IP PBX facilities. The routers selected for use in the Data Bank network (Cisco 2503) do not support this capability. Higher end devices such as the 2600 or 3600 family would have been required.

Data Bank elect to use the standalone IP telephones. These can be delivered to branch managers offices, plugged in and set up with minimal effort and inconvenience. They can also be removed or relocated should they prove ineffective at some locations.

In order to make IP telephony work, we need systems in the network that allow calls to be set up between IP telephones and also to interconnect the IP telephones to the conventional telephony network. These functions are known as:

- Gatekeeper. This is a server device that controls call set-up between IP telephones or between an IP telephone and the interconnection device. Each IP telephone has a conventional extension number. Users dial this—not an IP address, so the Gatekeeper has to do a translation function (similar to the role of a Domain Name Server). All new IP telephones have to register with the Gatekeeper, so that it can build an association between its IP address and extension number in a database. When an IP telephone places a call, it does this by passing a call request to the Gatekeeper. This is able to convert the required extension number to an IP address and pass an incoming call message to the destination, causing its bell to ring. If the call is answered, the receiving phone notifies the Gatekeeper, which in turn passes a success message (containing the destination IP phone IP address) to the calling phone. Thereafter, voice IP packets pass directly between the source and destination phone, without Gatekeeper involvement.

- Gateway. If the Gatekeeper cannot translate a destination phone number into an IP address, the destination must be on the conventional telephony network. The call is therefore passed to a Gateway. This is able to terminate the IP telephony traffic and convert it for onward transmission onto conventional telephony trunks.

Data Bank will use an IP PBX product which combines the Gatekeeper/Gateway functions. This will be located at the regional office and will interconnect to the regional office conventional telephony PBX system. The IP PBX is based on a PC server hardware platform, and is therefore significantly less expensive than conventional PBX equipment which tends to use custom hardware platforms. The IP PBX has the ability to connect conventional telephony handsets and convert them to IP telephony. Data Bank will use this capability to support a limited trial of IP telephony at the regional office sites. An outline design for the Data Bank IP Telephony trial is shown in Figure 10.19.

Logical design

Getting the IP telephony trial to work in a quality fashion will not be simple. The major issue Data Bank have to face is how to provide adequate Quality of Service over the data network for voice traffic, without compromising mission-critical business data applications. The three main demands for quality voice traffic are:

- Adequate bandwidth—lost packets will result in voice traffic 'breaking up'—similar to a poor quality cell phone connection.

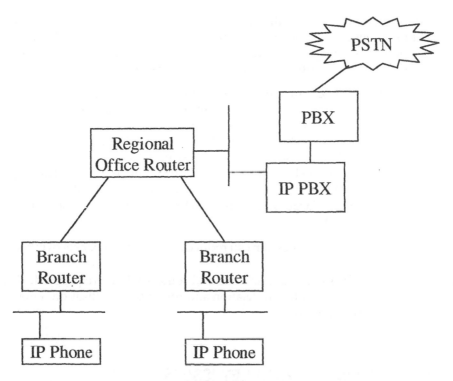

Figure 10.19   IP Telephony for Data Bankers

- Consistently low latency (delay). Traffic on IP networks suffer variable length delays, due to the statistically multiplexed nature of the service. The receiving station buffers packets before replay to overcome the variable delay. Human users are very sensitive to delay in voice calls and anything above 150 ms delay makes conversations increasingly difficult. Receive buffers therefore only have a limited depth, and any packets arriving excessively late cannot be used, so are discarded.

- The inevitably of delayed voice when sending over a packet switched network means that echo can be a significant problem. Voice over IP equipment selected should include echo cancellation technology to eliminate this problem.

To ensure adequate Quality of Service for voice, the following steps can be considered in the Data Bank network:

- Adequate bandwidth can be ensured by using a prioritisation process in the branch and regional office routers. This would be configured so as to give voice traffic a guaranteed share of the bandwidth. Selection of prioritisation scheme and its configuration must be done carefully, to ensure ample bandwidth remains available for business-critical data applications also. It may be necessary to experiment with several schemes to find the optimal configurations. For Cisco routers, the Cisco Waited Fair Queuing algorithm is a strong solution candidate.

- Voice packet delay must be minimised by reducing the size of packets sent on the network. If a voice sample has to wait for a 1500 byte Ethernet packet, it will be delayed by some 95 ms on the 128 kbps low speed branch sites, which is a significant proportion of a 150 ms delay budget. It will be necessary to set routers up to offer a reduced packet size (MTU) over the branch to regional office link, e.g. 512 bytes might be selected. This reduces delay for voice samples, but at the cost of fragmenting any large data traffic packets, increasing the amount of packet overhead.

- Using IP header compression can reduce the amount of packet overhead.

Careful monitoring of access links and routers will be required during the trial, to ensure that link utilisation and router CPU utilisation remains within reasonable bounds.

10.9   SUMMARY AND KEY ISSUES

In this section, we have presented a brief description of Data Bank and their networking requirements. The level of detail here is as might be found in a typical 'management overview'. In reality, this would introduce a more detailed statement of requirements often running to several hundred pages in length. We went on to provide outline design solutions for both current and anticipated future requirements.

The key issues to note from the section are that:

- We need to gain a considerable level of background knowledge of the customer and their requirements before design can begin. Specifically, you need to understand the business and business applications that the enterprise network are to support.

- There is not necessarily a 'correct' answer for the solution architecture. Many factors will influence choice of technology, including performance, price and customer preference. It is important to keep an open mind early on in the design process, and carry out top level 'thumbnail' design and pricing exercises using a range of technologies. A pros and cons analysis can be carried out and discussed with the customer. The designs presented above have been chosen to show a range of technologies typically used in today's customer networks.

- In carrying out network designs, it is essential to put at least as much effort into the logical design (i.e. how will it actually work?) as the physical design (what components does the network consist of?). Logical design considerations often have significant impacts on the physical design. Failure to properly consider logical design can lead to expensive network upgrades at a later stage (e.g. if insufficiently powerful routers were selected as OSPF Area Border Routers).

We have introduced potential new applications that Data Bank may wish to deploy. These show that the data network designer has considerable challenges going into the future, for example providing adequate security for Internet banking and Quality of Service for demanding multimedia applications such as voice over IP.

11

To Boldly Go . . .

It is a very sad thing that nowadays there is so little useless information

Oscar Wilde

If you diligently followed every step of the case study in the previous chapter (and worked through all of the open issues), you deserve a rest—and now is a good time to relax. The hard work is over and the time has come for some contemplation. The following may not have much technical rigour or academic bite but it does constitute an important part of understanding how and why networks evolve.

Let us take stock of what we know now. Basically, we have a fairly comprehensive understanding of what constitutes good design practice. Many of the principles we have explained in the body of the book will persist—the network designer will always have to attend to issues of performance, security and scalability. The way in which the principles are applied will change, though. Most noticeably, the technology used within a design moves on all of the time—it is axiomatic that technology will get bigger (and better) and speeds and volumes will grow. Moores law (which states that computing power doubles every 18 months) has held for over ten years. Less obvious is that the context in which a design is deployed will change. Although indirect, economic and social factors do have some impact on the design of enterprise networks.

In this last chapter we explore a little of the world to come. A degree of judgement is applied in doing this and we would not pretend that this chapter gives the same sort of hard facts and guidelines as earlier. Nonetheless, there are some trends that seem set to continue and events that should run to a logical conclusion.

11.1   SUPERCONNECTIVITY

Evidence of rapid advance in technology is all around us. The power of PCs on sale in the high street and the ever decreasing size of mobile phones are testament to this. Less obvious is the dramatic growth in bandwidth—the raw transmission capacity that carries the bits used by the PC and mobile phone across the globe.

A series of high capacity optical fibre links, each bigger than its predecessor, have led to a huge growth of available 64 kbps international circuits—over 50% per annum. And this huge capacity makes it feasible for even more people to be connected to some kind of network. Figure 11.1 illustrates this by charting the progress of intercontinental bandwidth that has been made available over the last ten years or so.

The dramatic rise in connectivity across the globe is nowhere more apparent than with the Internet. If the current rate of uptake carries on as it has over the last few years, it is not long until we are all surfing the night away.

The point is that networks are far from an exception; they are becoming the rule. The capacity that enables them exists and people are keen to become connected. When you have a universal fabric and people who are

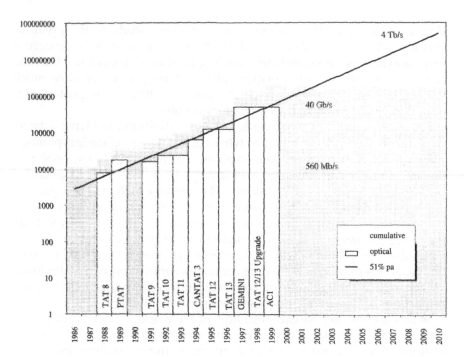

Figure 11.1   The Growth in World-wide Bandwidth

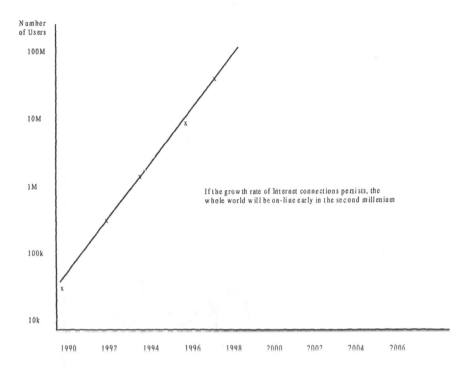

Figure 11.2 The Rise and Rise of the Internet

knowledgeable enough to exploit it, it comes as no surprise that networks play the same central role at work. The argument that networks will, increasingly, lie at the core of business is further supported when we look at the way in which organisations have changed over the years.

Figure 11.3 shows the trend away from self-contained organisations towards more distributed and federated enterprises. The technology needed to support the latter has to be agile and flexible at the same time as being secure and performant—the very issues already explained as key aspects of designing enterprise networks.

The underlying driver behind all this is that we all now tend to work with so much more complexity than previously. To cope with this you need the sort of high powered information tools that only an enterprise network can deliver. To illustrate just how much more data flows over greater areas than ever before, consider the quote at the start of the chapter and the somewhat startling illustration in Figure 11.4.

So, enterprise networks will undoubtedly become an indispensable part of every major organisation's infrastructure and competence in their design an important skill. Although significant, we will now leave that subject and set out some specific predictions to help set the context in which businesses are likely to operate. The next section makes a start on this with

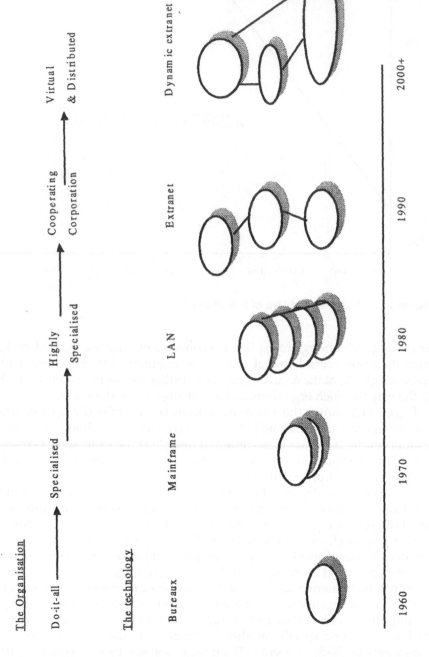

Figure 11.3 The Move Towards Federations

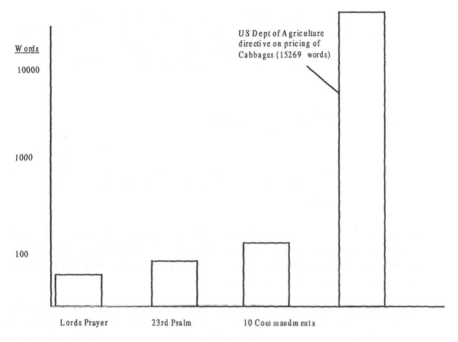

Figure 11.4   Infosurge

some technical futures. A more general set of predictions follow this as we consider some of the more significant economic, social and political factors that seem likely to shape the world in which we live and work.

11.2   SAFE BETS FOR THE FUTURE

Given that solemn predictions exist to make our descendants laugh, this section should be viewed as a series of 'author's prejudices', each presented in the form of a proposition with some evidence to back it up. Our view on technical futures is as follows.

Open systems will happen

The merging of the world of information technology and telecommunications, catalysed by the Internet, has started a move towards openness in technology systems. Computers are already becoming a personal commodity item. A greater penetration of computing devices into business (and the home) will be assured.

- Adherence to the World Trade Organisation Information Technology Agreement that removes tariffs on IT products by the year 2000 (which already has 40 countries representing 90% of world supply signed up) will result in continued reduction in the price of computers.

Distribution will be the norm

Distributed processing and openness have been goals of the Object Management Group and Open Group for some years. Present standards work is aimed at increasing the openness and portability of Object Request Broker-based solutions.

- Distributed processing standards will be well developed and open, hardware independent software (e.g. Java) will be the norm. Design and development of distributed systems will be a normal skill set of development engineers in the industry.

- A standard browser protocol capability will be agreed, forced by the market to allow user mobility and services of higher perceived customer value (like interactivity or personalised accounts accessible from anywhere).

IP will be everywhere

The advanced functions enabled by Internetworking Protocol (IP) based features now under development (e.g. Mobile-IP, differential quality of service, session handling, IP version 6, security mechanisms and multicast) will be the basis of most networked applications. Standard solutions for IP technologies will be deployed globally.

- Progress towards the standardisation of Customer Premise Equipment e-commerce technology will have been made. Standards for encryption, watermarking, authentication and signatures will be stable and deployed.

- Smart cards will not have significantly penetrated the market—less than 3% share of the North American automatic teller machine card market is predicted in 2002. There are presently a variety of 'card' technologies in use and three main standards for the operating system exist for them (MasterCard™, 'Multos' Microsofts Smart Card for Windows and Visa™, 'JavaCard').

- Costs to merchants of supporting smart card technologies of $120–250 per terminal require high usage.

- Deregulation (balanced by copyright and security) will have allowed open and cost-effective access use of e-commerce technology tools internationally.

Greater sophistication can be expected in this area. For instance, Directory Enabled Networks (DEN) will influence the design of future networks. In much the same way the extensive markup language (XML) seems likely to provide a basis for smarter applications.

Big pipes to the home

Access bandwidth technologies, and the associated core network standards that are needed (and implied), will have matured and regulation and standards will have enabled both global interconnect as well as some degree of open interfaces onto network capabilities.

- Access (and network) technologies are still 'low' in bandwidth. ISDN is prevalent within Europe (over 30 million lines) with xDSL starting to pick up.
- Access technology solutions in North America will be dominated by xDSL and cable modems. The Asia-Pacific market will be dominated (50%) by radio access lines.
- Within the European core network Asynchronous Transfer Mode (ATM) technology will be important but with a significant and growing use of IP as the networking protocol. In North America IP and related technologies will be dominant.
- Universal Mobile Telecommunications Services (UMTS) technology will be maturing and deployed on a small scale. With economic factors neutral and societal forces strongly for the paradigm, the slow development will have been due to political regulation (such as spectrum allocation problems).
- GSM (Global Systems for Mobile) and CDMA (Code Division Multiple Access) systems proliferate along with mobile-IP infrastructure. Predictions suggest that mobile infrastructure in the west will be dominated by Digital GSM (30%) and CDMA (10%).

There is considerable impact in these observations on the role and job of the designer. With such technical diversity and flux in network technology it is vital to find time for continuous education and training, learning about new technologies as they emerge, established ones as they develop and keeping in touch with the wider changes in business and society which drive their uptake. Some of the ways in which designers can keep up to date are covered in Chapter 8. In addition to this, most of the professional engineering bodies (IEE, IEEE etc.) have a strong emphasis on lifelong learning.

˙ 11.3   A CHANGING WORLD

In the coming years, the whole of the communications market, and the environment in which it operates is likely to be quite different from today. For any organisation that trades on information and its associated technology, it is important to understand what such a world may be like and what change means. This is not confined to the technology predictions outlined above, although changes here are very important. Political, economic and society aspects are also important as they structure the marketplace in which the players do business.

This section presents analysis and information on present trends that can be extrapolated to describe views of the future world and the impact that they may have on the roles described. In each area, some key propositions are put forward along with some published supporting information and general trends.

The economic world

The background factors and trends considered are as follows. The migration of power from state to companies will have continued and the power of the global economy will be significantly stronger than it is today.

- Already over half out of the top 100 economic entities are companies not states.

- If the world is considered with the wealth of Italy today as a unit, then the USA is eight Italys, Western Europe another eight, and Japan a further five. The rest of Asia, South America and Russia constitute the remaining five.

- Five Italy's are 'spent' by companies each year purchasing other foreign companies, buying into that country's market.

- Today is the age of skills, and the service economy (the GDP of a western developed nation is divided into approximately 5% agriculture, 20–30% manufacturing and the remaining 60–70% the service sector).

- The USA has seen a 23% increase in the number of jobs since 1980 which has all been in the private sector. The growth in small and medium sized enterprises (< 100 workers) has been twice the average.

- The developed countries collective productivity (the ability to utilise technology to integrate the value and supply chains together) has been and will be the key that enables them to retain wealth, and continue to grow their economies as other nations develop (e.g. China).

Competition will be most vigorous in the provision of tailored services to end consumers (mass customisation). Value to the end customer and cost of production are the key considerations. Potential markets for service providers will come and go as the needs and aspirations of the many individualised consumer types change.

- The service economy will be segmented across all sectors into small and medium enterprise (SME) niche providers able to respond rapidly to the frequent shifts and developments in their markets.

- These SMEs are either virtual (structured from networked 'self-employed' skilled individuals) or very localised geographically but will be able, through the development of open networked technology, to offer their service to the global market.

- They will, however, be very sensitive to market change and less stable as a result—the services that small and medium sized enterprises offer change as rapidly—and employment will be dependent on people's skills and experience.

The world of infrastructure provision will be less dynamic and competitive. As a consequence of the huge capacity illustrated in the first section, price and cost will be the key concerns. Bandwidth exchanges will have become quite well developed and capacity will be a commodity item that is traded at a published, and fluctuating, market price.

- Competition in the provision of network or access infrastructure occurs when technology opportunities arise.

- The market has drastically reduced the profitability of the connectivity supply business and only a few, well established and stable players exist as a result. Large-scale geographic networks are operated by large stable players or alliances.

- The access environment may be a little more complex. For example, housing associations and developers may opt to install communications infrastructure on estates that they then own and maintain. They then sell connection to this 'valuable' access infrastructure to the players offering networks.

One important trend from the above list that affects all individuals is the growing move towards a world dominated by a 'knowledge-based economy'. Countries with high labour costs (e.g. the US) or a relative lack of natural resources (Sweden, Switzerland or Japan) have long realised that technical skill and innovation are their key to maintaining a competitive edge in the global marketplace. They structure their industries accordingly, so much so that up to half of the fastest-growing US companies are 'knowledge companies'. They sell the knowledge and skills of their employees rather than manufacturing products or providing services.

Networking tools such as the Internet greatly facilitate the ability to gain access to knowledge spread and share it. Of course, with an increasing realisation of the value of knowledge, the evolution of network security techniques will have an important role to play ensuring that only those entitled to access knowledge can gain it.

A changing society

Ultimately, it is people who determine whether technology is exploited or rejected. Today, individuals are taking on more and more responsibility for their futures (in terms of attending to their own healthcare, pensions, insurance etc.) and the service sector marketing is starting to become 'personalisation' dominated.

Traditional community-based links will have become supplemented (or supplanted) by links chosen by the individual. Work, hobbies, affiliations etc., which may not be so local, will drive these links. The 'family' is likely to have a very wide meaning covering many forms, with interest networks (friends, hobbies etc.) becoming important.

- Local interest groups will also be encouraged by the existence of community networks supported and encouraged by government. Home and work will have become less distinct, teleworking will have become more prevalent as technology.

The concept of communications services will have changed significantly. Little distinction in value will be made between voice, video and data. Users will have begun to accept that quality is balanced with price and that premiums will apply to personalisation of factors (such as reliability or additional requirements) that are beyond the basic service package. The present trends towards regular, informal, and instant communication interaction will become the expected norm.

- To support social interaction communications will be more network-based, increasingly in 'electronic' form both for written and spoken interaction.

- Instancy will be vital, people will not want to wait long; cellphones are instantly enabled for mobile telephony services whereas more traditional voice 'services' have long lead times at present.

- Through experience with the Internet users will have got used to only waiting a few seconds/minutes for the information they have requested. They will expect the power of the technology they have invested in (computers etc.) to be enough for any task that they undertake.

Attitudes to computing technology will be very different to that adopted with the telephone. Unlike the telephone, which has been around for many years and has just reached around 95% consumer penetration in many of the more developed countries (although, in some third world countries penetration is as low as 0.2%!), the computer becomes part of everyday life from school age.

- Studies indicate that only about 60% of the population are comfortable with the telephone but over 90% of UK children (aged 11–18) in schools are confident in using computers/PCs.

- In addition, about 70% of parents, regardless of social class, felt it was critical that their children learn to use a PC. Assuming that those aged 18–25 today had a similar experience, by 2005 one-third of the UK working population will be computer aware and confident in their use.

- People across all age and social groups are already of the opinion that computers are an essential or important in their working lives. In 1997, 53% of all people felt computers essential, with 60–70% in the higher earning groups.

- The PC is penetrating into all sections of society, around 30% of people aged between 16 and 59 had a PC at home in 1997, a similar figure to that in 1992 but the machines were more powerful. For the higher earning groups, between 55–65% own home PCs. Of the smaller firms, 33% have portable laptop PCs. Ownership seems to be shifting emphasis away from games and entertainment and towards education, skills and training, echoing the socio-political emphasis on individual responsibility (the risk society).

- As with the culture of the computing peripherals, the young workforce consider the mobile phone to be the norm. They have grown up with the telephone and have no fear of mobile technology. Already the penetration into AB social classes is high with 43% ownership of mobile phones in 1996 and 68% of small firms having a mobile phone.

Society's approach to payment methods, increasingly linked to trust in both technology and companies, is also starting to change. People are beginning to show some trust, and will start to use, electronic methods of payment as the risks to them (fraud, error etc.) have been minimised; familiarity grows and a track record is established.

- There are an increasing percentage of people who are happy to use the telephone as a purchasing channel. Studies suggest a recent increase of

10–20% for a wide range of goods/services (from insurance to shopping to holidays).

- Many people are happy to use credit cards on the Internet. American Express estimate that between \$4–6bn was purchased using credit cards in 1997 compared to \$68bn private consumption in the US.

- New payment technologies like smart cards need to identify the value and benefit they give to the user. The Mobil™ 'Speedpass' is an example showing that users will take up the technology if the incentive is right—it makes life simpler and does not cost a lot. The lack of widespread ability to use electronic cash is the present hurdle to adoption by society. Initiatives, such as the electronic payment system for London Transport that was recently in the news will stimulate market adoption by generating a large captive, high density user base.

- Business to business commerce, e-business, can be efficient, as it can remove a number of intermediaries from the supply chain. By 2002 it is predicted that the businesses–business interaction on the Internet will rise to generate tens of Billions in revenue (Norris, West and Gaugan).

- There is strong evidence supporting the increased personal mobility of people. By 2005 a third of the workforce will expect to use mobile technology at work and will increasingly own and use such equipment in their social time.

- Figures from the UK department of transport suggest that there will be a 20–30% growth in air travel compared to 1997. The growth is expected for both business travel and tourism in general: UK air business departures will reach 9.7m short haul and 4.6m long haul and arrivals will reach 10.7m and 5.2m by 2001.

- Mobile phone connections are also expected to rise rapidly, and by 2003 it is predicted that there will be 23.6 million mobile connections in the UK. The rest of Europe will have 131 million mobile connections and the situation is similar in North America and Asia-Pacific. Pay-as-you-go marketing has done much to make the mobile phone a commodity item.

Global politics

The increasing power of some businesses (especially multi-national corporations) and the dominance of the global economy has diminished the role of government and reduced its influence. Government will become even more involved with regulation and control rather than leadership.

- Increased globalisation will focus government at a world level (supranational governments) and support regulation on a global scale (World Trade Organisation, World Intellectual Property Organisation, International Monetary Fund are existing examples). Countries are now just members of these global government bodies, with ministers acting as representatives, and have to agree to consensus views.

- The conflict between company and people's interests continues but a nation's representation generally reflect the interests of companies.

- People have little respect for politicians and the process of representational government. Politicians are no better educated and the media has given everyone a view. There will be an increasing wish by individuals to have more say, and representational government is now accompanied by a series of referenda.

Governance will therefore have moved to be more 'local', focused on supporting the natural needs of individuals. To support this community of interest networks (COINs) become the norm, government (local and national) being a key player and this will represent a new market opportunity for players in the telecommunications industry.

- The availability of affordable computing tools in the home will help local government, via these community networks, to communicate better with their constituents.

- The move to local government will be hastened by society's recognition that people elsewhere have an effect on their local circumstances (wealth, environment).

The enhancement of the regulatory role of government and its localisation will result in increased political stability within nations. Globally, the growing economy brings political stability, there is less scope for different government policy between left and right. The destructive forces of nationalism will not have developed as people recognise that stability requires co-operation and sharing (for example, the single European currency, the Euro, is aimed at stabilising the European economy).

Privatisation of telecommunications will have occurred in all major markets around the world. The way the UK and European market is regulated has changed significantly from the picture of the late 1990s. The regulation will have forced significant openness.

- Access by competitors to facilities within infrastructure such as switch interfaces, main distribution frames, and cross-connection equipment will be normal.

- The Open Network Provision and Convergence European Union (EU) regulation will be in place.

- Developing nations will adopt similar (identical) policies to the EU in their own new and open markets believing that growth in their economy is best enabled by shared infrastructure and opening core systems.

- Transmission capacity will be traded as a commodity with a price set by market demand. The provision of infrastructure will be separated in regulation terms from the delivery of service, which is regulated only by market practices (fair trading etc.). The role of the telecommunications regulator will be radically different from that of today.

Privatisation will spread. The role of the state in organising, investing and managing the utility sector and other 'service' activities is all but gone. Government regulates and manages those parts of the economy that cannot be privatised. Privatisation programmes only last for a short period as society loses interest and so by the year 2005 the western economies will be fully privatised.

- There are presently privatisation programmes in 100 countries throughout the globe expected to generate $200bn.

- In the UK the subsidy paid out by government has been turned into income for Treasury. A £1.5bn subsidy and loan finance to 33 companies in 1979 turned to an £8bn contribution to the Treasury in 1987.

These political factors have a different effect on the various players in the communications industry. For instance, service providers benefit from the ability for people to chose (democracy) and the absence or presence of any political hurdles to entering markets (nationalism). Providers of network capacity are affected more by regulation and by whether their home market is open to fair competition, or not (state ownership).

11.4 SO WHAT?

J. B. Priestly once observed that prediction only serves to make our descendants laugh. Given this, it would be unwise to draw too many conclusions from our analysis of the world, as presented above. In any case, the best way to predict the future is to invent it and this is probably more the case in the world of networks than many other areas. To close then, we make one final observation.

There are several ways that nations cede power to supranational bodies. National governments are ceding power upwards to supranational bodies and also downwards to regional and local decision makers (e.g. regional assemblies). There is a general principle at work here—that of subsidiary, where decisions are deemed most effective when taken at the lowest

possible level, reflecting only the grouping of people affected directly by
the decision.

Clearly, with decisions, laws and regulations being made at local, national
and supranational level, the use of networks is essential, both in helping
make the decisions in the first place and in communicating the outcome of
the decision process. The White House and the European Commission, for
instance, both have excellent web sites and appear to see this as a key
method of communicating with citizens.

The main focus of the book has been on the engineering aspects of
enterprise networks. But the remit of the designer can go much wider than
that. A more complete picture emerges when the designer is aware that
they are building something that works in a complex environment. That's
the real art of design and that's what the book is all about.

BIBLIOGRAPHY

Barnes, I. (1991) Post Fordist People. *Futures*, November.

Cooke, P. *et al.* (1993) *Towards Local Globalisation*. UCL Press.

Headlines 2000 (1989). A report of the Cookham Group. Hay Management
    Consultants.

Drucker, P. (1993) *Post Capitalist Society*. Butterworth Heineman.

Gore, A. (1991) Infrastructure for the global village. *Scientific American*,
    September.

Guilder, G. (1993) When bandwidth is free. *Wired*, September/October.

Hague, D. (1991) *Beyond Universities. A New Republic of the Intellect*, Institute
    of Economic Affairs, London.

Handy, C. (1996) *The Empty Raincoat*. Arrow.

Morton, M. (1991) *The Corporation of the 1990s*. Information Technology
    and Organisation Transformation Oxford University Press, New York.

Naisbitt, J. (1994) *Global Paradox*. Nicholas Brealey.

Norris, M., West, S. and Gaugan, K. (2000) *E-Business Essentials*. John
    Wiley & Sons.

Editorial (1999) The knowledge based economy. *Nature Magazine*, January.

Ohmae, K. (1990) *The Borderless World*. HarperCollins.

Porter, M. E. (1986) *Competition in Global Industries*. Harvard Business
    School Press.

Reich, R. (1991) *The Work of Nations*. Simon & Schuster.

Appendix

A Quality Approach to Network Design

Good judgement comes from experience and experience comes from bad judgement

Fred Brooks

The history of quality management can be traced back as far as the Pharaohs in Egypt, when funerary goods had to be approved by the superintendent of the Necropolos and had to bear his mark. This early example illustrates the two key principles of quality management—that suitability for purpose should be checked and that a record is kept of that check.

Companies throughout the world are increasingly implementing quality management systems. Not only do they encapsulate good practice, they also put it into a formal framework where it can be audited and measured. For our purposes the relevant guidance is contained in the ISO 9001 standard (ISO 1994), and many will seek to have their quality system formally audited and certified to this standard. Such companies (and even many that are not yet certified) will look to their suppliers to also have ISO 9001 certification. (Readers familiar with earlier versions of ISO 9001 should be aware that the 1994 version of this standard includes significant new requirements in the area of design. These requirements are explained in this section.) The purpose of this section is to explain to the enterprise network designer the principles of this standard, specifically the approach that will be required to the design task as a result of it.

The ISO 9001 standard centres upon the relationship between a customer and a supplier, who contracts with the customer to supply goods or services. In the context of this book, the goods and services will be key components of the enterprise network.

The standard works by requiring formally documented processes to be used at all stages of product supply. These processes will include a range of quality checks, such as document reviews, configuration and audit plans, product testing and corrective action plans. In all, there are 20 requirements that need to be satisfied to meet the standard. The supplier is expected to document the result of such testing, and retain this documentation as evidence that the quality management system is being followed (Norris, Rigby and Payne 1993).

In practice, adoption of a quality management standard sets the supplier of a network a clear objective of achieving customer satisfaction. The measure of achievement is the degree of prevention of non-conformity to the customers specified requirements at every stage, beginning with the design process. Therefore, before we can look at the implications for design, it is necessary to review the implications for the input to the design process (i.e. Customer Requirements). As with any process, the rule 'Garbage In—Garbage Out' applies!

The 'quality assurance' view of the network design lifecycle is shown in Figure A.1. This can be refined to show a series of quality gates between separate stages.

Installing a quality management system need not be particularly onerous. After all, the whole essence of such a scheme is to be able to say what you do, do what you say, and to prove it. The key aspects of of meeting ISO 9001 are to get senior management commitment, to establish clear responsibilities, to identify key activities/measures and to establish regular audit and review.

The essential aspects of network design are reflected in the main chapters of the this book and this gives us a direct lead in to the areas that need to be covered by the quality management system, as below.

A.1   CUSTOMER REQUIREMENTS

Pivotal to the whole ISO 9001 is the meeting of 'Requirements prescribed by the purchaser and agreed by the supplier in a contract for products or services'. Before design work can begin, therefore, agreed requirements are necessary, either in the form of a Statement of Requirement (at the contract bidding stage) or Contract (post contract award). The ISO 9001 prescribes a formal contract review process before post win design begins.

ISO 9001 Network Design Process Control Model

Design Management Tasks

Agreed Customer Requirements → Contract Review

Plan

Other Requirements → Design Input

Review

Resource

Design Process

Reviews & Verification

Design Change

Design Output

Final Review

Verification

Network In Use ← Implement Network

Validation

A.1   Network Design Control Process Model

A.2   CONTRACT REVIEW

ISO 9001 requires that a supplier must establish and maintain documented procedures for contract review and for the co-ordination of these activities. Records must be kept to demonstrate that reviews have been carried out, and to record the outcome. During contract negotiation, there is a need to review the contract requirements in a systematic and documented way. The contract must be reviewed by the supplier to ensure:

- Customer requirements are adequately defined, understood, agreed and documented.

- Differences between what the customer requires and what the supplier can supply are resolved with the customer and agreements documented.

- The supplier has the capability to meet the contract requirements in terms of resources, organisation and facilities.

A.3   DESIGN INPUT

Input to the post win design process consists of:

- the reviewed contract;
- any relevant regulatory requirements;
- any relevant international standards;
- supplier's design rules and standards;
- manufacturer's information on components to be used.

A clear record of what the design input is should be maintained and the selection reviewed for adequacy. Any incomplete, ambiguous or conflicting requirements must be resolved with those imposing the requirements. Collectively, this input may be referred to as the Design Specification or Design Brief.

ISO 9001 notes that the design specification will seldom be complete at the beginning of design work. Specifications and requirements can frequently only be refined as part of the design process.

A.4   DESIGN PROCESS

The purpose of the design process is to translate the agreed customer requirements, in a systematic and controlled way, into a specification which defines and facilitates implementation of the service to be provided.

ISO 9001 requires that the design process is clearly documented. The procedures need to include mechanisms to:

- Control and verify the design work, in order to ensure that the specified requirements are met. This will include documenting organisational structure, responsibility and authority.

- Provide a design plan for the design activities, and ensure that the plan is updated as the design evolves. Plans must include the identification of responsibility and resources for each design activity, timing of activities, checkpoints to monitor progress and the scheduling of reviews and verification. We must envisage design not as a one off process, but a continual commitment during the life of a contract, with ongoing change management work, and periodic review. This can be reflected in the design plan.

- Ensure that suitably qualified personnel equipped with adequate resources, are assigned to do the work.

- Document organisational and technical interfaces to other teams who input to the design process.

A.5   DESIGN OUTPUT

Design output must be documented, and expressed in such a way that it can be readily verified and validated against the design input requirements. The output must:

- meet all design input requirements;

- contain references to suitable acceptance criteria, which will allow the design to be validated on implementation.

All design output must be reviewed before release, and as with all ISO 9001 documentation, it must be properly version controlled, ensuring that end users (e.g. the network implementation and network management teams) have access to the latest, controlled versions.

A.6   DESIGN REVIEW

At each stage of the design, formal documented reviews are to be planned and conducted. Participants in the review should include representatives of all teams concerned with the stage under review, as well as specialist consultants as necessary. Results of reviews are to be documented and retained. The reviews should be performed by people independent of the actual design process.

The objectives of the review is to confirm that:

- The document under review conforms to required document standards.

- The design is free of major errors.

- The documented design is viable.

- The design will meet specified customer requirements.

A final design review, prior to implementation, is mandatory.

A.7   DESIGN VERIFICATION

In addition to design review, which focuses on design documents, verification is required. Verification should be an integral part of design work, but formal reviews of the verification work are required at stages throughout the design, with a final verification prior to implementation. The purpose is to confirm that the design, when realised, satisfies the design inputs. Work may include:

- performing alternative calculations;

- comparison of the new design with similar proven solutions;

- undertaking lab tests and demonstrations.

A.8   DESIGN VALIDATION

This is done after verification and during delivery of the customer's network. Again, documented procedures and controls will be required. The purpose is to:

- Confirm that the network meets the customer requirements, by carrying out an agreed programme of acceptance tests, conducted under defined conditions.

- Auditing the delivered network to ensure that it conforms with the description in the contract and design documentation.

Several stages of validation may be required if the network delivery is phased (e.g. trial, pilot, roll-out).

A.9   NON CONFORMANCE

In the case that any non conformance is found during review, verification or validation, one of the following actions should be taken:

- The design/implementation should be reworked to meet the customer requirements.

- A concession should be agreed with the customer, and a programme of actions agreed to resolve the issue in the longer term.

An issue management system will be needed to record and manage these and other design issues. ISO 9001 recommends the corrective action procedure for handling non conformance in delivered product. This is a strictly controlled problem resolution mechanism, with clearly defined time constraints, formal tracking and documented escalation procedures.

A.10   DESIGN CHANGE CONTROL

Design work is under formal change control. Change requests to the design (these are normally initiated through changes to the design specification) must be prepared and recorded. The design documentation should be version managed, and management decisions will be needed as to how to incorporate changes into different releases of the design.

All change requests should be reviewed and approved by authorised personnel before being fed into the design process. Before permitting a change, aspects to consider include:

- Will the service still meet the customers requirements?

- How must the design specification documentation be changed to accommodate the change?

- Are there any 'knock-on' effects of the change?

- Does the change create problems for implementation or management of the network?

- What impact will there be on verification and validation procedures?

- What are the cost and timescale implications? Are these agreed by the customer?

A.11   MANAGEMENT RESPONSIBILITIES TOWARD THE DESIGN PROCEDURE

Management have the responsibility to ensure that the design process and related procedures are in conformance with the supplier's corporate Quality Management System.

Management are responsible for seeing that necessary procedures for design are developed, documented and followed. The procedures must be periodically reviewed, according to documented procedures. The objective of review is to ensure that the procedures are effective and reflect actual practice.

Management are responsible for collecting, indexing and maintaining quality records, demonstrating that procedures are being carried out.

Additionally, management should identify, implement and monitor any necessary measurements of the design process to ensure the process is under control, and identify areas for improvement.

Management are responsible for seeing that design work is properly resourced with people, who have the appropriate skills and training to do the necessary work. It is necessary for management to define and document the responsibilities and authority of these people, ensuring that they have the authority and organisational freedom to:

- Initiate actions to prevent non conformance of design to customer requirements.

- Identify and record any problems and issues relating to the design, its performance or the governing quality system.

- Initiate or provide solutions to problems through appropriate channels.

- Verify the implementation of solutions.

- Control further implementation until serious problems are resolved.

- Management should identify and procure any non manpower resources necessary to the design process, e.g. design tools and test facilities.

A.12   CHECKLIST: A DESIGN PLAN

The following table presents an outline design plan, which can be used to track key deliverables and milestones for the process of network design that we have discussed previously. The design process is iterative, and will move from version to version as it evolves. It is suggested that a separate plan be established for each version.

A.13   GUIDELINES

By way of suggestion for the checks that should be made through the design process, here are a few of the less obvious ones:
Project Management

- The estimation process and its results should be documented along with any other project assumptions and risks.

- Project risks should be assessed throughout and a variety of viewpoints should be sought when they are assembled.

Table A.1

Outline Design Plan for Design Version X.Y	Planned Date	Achieved Date
Contract review complete (requirements understood)		
Change and version control agreed		
Other design input documented and reviewed		
Design documented and reviewed		
Design verified		
Acceptance test spec prepared and agreed		
Testbed acceptance testing complete Results reviewed		
Design validated (Acceptance testing on pilot network complete and results reviewed)		
First post implementation design review complete		
Second review		
Third review etc.		

Requirements Capture

- As much attention should be paid to job and process design as to network equipment and computer systems.

- A range of people need to be consulted to ensure that the requirements gathered cover all of the relevant viewpoints.

Verification, Validation and Test

- It is likely that standard tests will be required over and again, so establishing an environment for regression testing is usually worthwhile.

- Gathering metrics on defects gives useful guidance on when to stop testing and on where improvements are needed.

Measurement

- Combinations of metrics are often more telling than a single metric.
- An approximate answer to the right question is much better than a precise answer to the wrong one.

Configuration Management

- Baseline records should be kept even if the project as a whole is replanned. It is important to know variance from original plans.
- A naming convention should be established as soon as possible.

Maintenance/Release

- a schedule of all releases (major and minor) needs to be maintained. It is virtually impossible to maintain a network without good information on its structure and set up.

A.14 SUMMARY AND KEY ISSUES

This appendix has aimed to present the key requirements of a quality management system for the design of enterprise networks. The key objective of such a system is to keep the customer satisfied—by ensuring their requirements are fully met. The main points of the quality system are:

- Ensuring that the customer requirements are fully agreed, documented up front and kept in step as work proceeds.
- Carrying out the design work in a consistent manner, using documented procedures, with auditable evidence that the procedures are conformed to.
- Checking of conformance at all stages through review, verification and validation.

The key issue with any quality system is that they can be time consuming and expensive to set up and run. A supplier may be forgiven for thinking that such a system will add to their tender costs and reduce the efficiency of expensive and scarce network design personnel. A counter argument to this is that:

- Large enterprises are increasingly requiring suppliers to be ISO 9001 certified. The costs of *not* having such a quality system can be significant loss of business.

- The cost of failure is invariably higher than the cost of the quality system. Imagine a 1000 branch router network has just been completed. As traffic builds up, branches start to fail. Investigation reveals the all too common problem—not enough buffer memory in the routers. The customer is screaming for heads on platters, and the supplier is forced to upgrade all routers from 4 Mbytes to 16 Mbytes of memory. This takes 500 days to complete, with half of that done by costly contractors brought in to help. What if the quality system had forced the designer to verify his assumption that 4 Mbytes was enough by a simple lab test? It is possible that this problem may have been eliminated at the cost of one day's effort. A small investment in eliminating problems at design time can save fortunes after implementation.

One final point: validation is used to demonstrate that what we have delivered meets the customer's requirements—and naturally, occurs toward the end of the design process. We must bear in mind the possibility that the designer has not correctly understood the customer's requirement, or indeed that the customer has not been able to adequately articulate the requirement. Clearly, if this only comes to light at validation time, there can be much expense and delay in correcting the problem. To avoid this, it is essential that the customer is involved in design reviews throughout the design process—the earlier a problem is detected, the cheaper to fix.

BIBLIOGRAPHY

Anttalainen, T. (1998) *Telecommunications Network Engineering.* Artech House.
International Standard ISO 9000-3 (1994) *Quality Management and Quality Assurance Standards,* part 3.
Norris, M., Rigby, P. and Payne, M. (1993) *The Healthy Software Project—a Guide to Successful Development and Management.* John Wiley & Sons.
Ward, E. (1998) *World Class Telecommunications Service Development.* Artech House.

Glossary

Tyranny begins with a corruption of the language

W. H. Auden

It is probably apparent through this text that enterprise networks are complex and fast evolving entities. They are driven by, and draw on, a number of technical specialisms—software, local and wide area networking, computing and information technology to name but a few. Each of these areas has its own associated vocabulary. None is easy to follow and all grow ever more difficult for the non-specialist.

In particular, the convergence of computing and telecommunications has led to a broad and often confusing set of terms that are assumed by many in the profession to be widely known. We list here here many of the key abbreviations and concepts in common use these days. Some of the terms have been explained in the main text, many are not. In either case the aim is to clarify some of the more complex ideas by giving some idea of their context and application.

Abstraction	A representation of something that contains less information than the something itself. For example, a data abstraction provides information about some reference in the outside world without indicating how the data is represented in the computer.
Abstract	Containing less information than reality.
Access control method	A methodology of distinguishing between the different LAN technologies. By regulating each workstation's physical access to the transmission medium, it directs traffic round the network and determines the order in which nodes gain access so that each user obtains an efficient service. Access methods include Token Ring, Arcnet, FDDI and Carrier Sense Multiple Access with Collision Detection (CSMA/CD), a system employed by Ethernet.

Address	A common term used both in computers and data communication designating the destination or origination of data or terminal equipment in the transmission of data.
ADSL	Asynchronous Digital Subscriber Loop. A sophisticated transmission technique that allows high speed communication (up 2Mbps in one direction) over local telephone lines.
Aggregate	The total bandwidth of a multiplexed bit stream channel, expressed as bits per second.
Algorithm	A group of defined rules or processes for solving a problem. This might be a mathematical procedure enabling a problem to be solved in a definitive number of steps. A precise set of instructions for carrying out some computation (e.g. the algorithm for calculating an employee's take-home pay).
ANSA	Advanced Networked Systems Architecture, a research group established in Cambridge UK in 1984 that has had a major influence on the design of distributed processing systems. The ANSA reference manual is the closest thing to an engineer's handbook in this area.
Analogue transmission	Transmission of a continuously variable signal as opposed to discretely variable signal. Telephony networks have traditionally been analogue.
API	Application Program Interface—software designed to make a computer's facilities accessible to an application program. All operating systems and network operating systems have APIs. In a networking environment it is essential that various machines' APIs are compatible, otherwise programs would be exclusive to the machines in which they reside.
APPC	Advanced Program to Program Communication—an application program interface developed by IBM. Its original function was in mainframe environments enabling different programs on different machines to communicate. As the name suggests the two programs talk to each other as equals using APPC as an interface designed to ensure that different machines on the network talk to each other.
APPC/PC	A version of the Advanced Program to Program Communications (APPC) developed by IBM to run a PC based Token Ring network.
AppleTalk	OSI-compliant protocols that are media independent and able to run on Ethernet, Token Ring and LocalTalk. LocalTalk is Apple Computers' proprietary cabling system for connecting PCs, Macintoshes and peripherals and uses CSMA/CA access method.
Application	The user task performed by a computer (such as making a hotel reservation, processing a company's accounts or analysing market-research data).
Application program	A series of computer instructions or a program which when executed performs a task directly associated with an application such as spreadsheets, word processing, database management.
Applications software	The software used to carry out the applications task.

Architecture	When applied to computer and communication systems, it denotes the logical structure or organisation of the system and defines its functions, interfaces, data and procedures. In practice, architecture is not one thing but a set of views used to control or understand complex systems.
Artificial Intelligence (AI)	Applications that would appear to show intelligence if they were carried out by a human being.
Asynchronous data transmission	A data transmission in which receiver and transmission transmitter clocks are not synchronised, each character (word/data block) is preceded by a start bit and terminated by one or more stop bits, which are used at the receiver for synchronisation.
Asynchronous Transfer Mode (ATM)	ATM is a standard for high speed fixed size packet communications. It provides a basis for multi-service networks—those capable of carrying voice, video, text etc.
Automation	Systems that can operate with little or no human intervention. It is easiest to automate simple mechanical processes, hardest to automate those tasks needing common sense, creative ability, judgement or initiative in unprecedented situations.
B Channel	The ISDN term used to describe the standard 64 kbit/s communications channel.
Bandwidth	The difference between the highest and lowest sinusoidal frequency signals that can be transmitted by a communications channel, it determines maximum information carrying capacity of the channel.
Basic rate interface	An ISDN term that describes the two interfaces, 64 kbits/s transmission links and a 16 kbit/s signalling channel, referred to as bearer links and the delta channel. Also see ISDN.
Batch processing	In data processing or data communications, an operation where related items are grouped together and transmitted for common processing.
Beta test	Commonly used term to denote the system that is released for field use (by a fortunate set of trusted customers) but that is not completely tested.
BGP	Border Gateway Protocol. This is the protocol used in TCP/IP networks for routing between different domains.
Blocking	A situation when a path or connection is not available because all of those available are busy. Blocking is a phenomenon of circuit switched networks, where the designer trades off concentration against throughput. Most public switched networks are designed with sufficient resources to allow users to gain access virtually all the time, without being blocked by other users.
Brooks law	This is the maxim that adding programmers to a project that is late will cause it to be further delayed. Brooks explains the law in his 1975 book *The Mythical Man-month* which draws on his experience of the IBM OS/360 project. The law is as valid now as it was in the 1970s.

Bridge	A device or technique used to match circuits, thereby minimising any transmission impairment. Most commonly used to connect two segments of a local area network together.
Bug	An error in a program or fault in the equipment. A hangover from the days when an insect caught in an early electro-mechanical computer caused it to fail.
CASE	Common Application Service Element—a collection of protocol entities forming part of the application layer that are used for providing common services.
Core Delay	This is the delay experienced across a carrier network. An aggregation of link, switching and queuing between the designated entry and exit points.
CCITT	Consultative Committee of the International Telegraph and Telephone. Until the early 1990s, a key standards making body for public network operators. Superseded by ITU/T.
CEN/ CENELEC	The two official European bodies responsible for standard setting, subsets of the members of the International Standards Organisation (ISO). The main thrust of their work is functional standards for OSI related technologies.
CEPT	The European Conference of Posts and Telecommunications—an association of European PTTs and network operators from 18 countries. It is the sister organisation to CEN/CENELEC.
CIR	Committed Information Rate. In Frame Relay, this is the guarantee level of throughput.
Circuit	An electrical path between two points generally made up with a number of discrete components.
Circuit switching	The method of communications where a continuous path is first established by switching (making connections) and then using this path for the duration of the transmissions. Circuit switching is used in telephone networks and some newer digital data networks.
Client	An object which is participating in an interaction with another object, and is taking the role of requesting (and receiving) the required service.
Client–server	The division of an application into two parts, where one acts as the 'client' (by requesting a service) and the other acts as the 'server' (by providing the service). The rationale behind client–server computing is to exploit the local desktop processing power leaving the server to govern the centrally held information. This should not be confused with PCs holding their own files on a LAN, as here the client or PC is carrying out its own application tasks.
Connection oriented	When information is exchanged over a fixed link with predictable characteristics. The public switched telephone network exemplifies this type of network.
Connectionless	When information is dynamically routed across a network in self contained units. X.25 is a widely known connectionless service. This type of network is characterised by possible loss, delay or reordering

	of information. It is the user who has to implement the end-to-end protocols to reorder, resequence and recover.
CMIP/CMIS	Common Management Information Protocol/Service. A standard developed by the OSI to allow systems to be remotely managed.
Code	A computer program expressed in the machine language of the computer on which it will be executed, i.e. the executable form of a program. More generally, a program expressed in representation that requires only trivial changes to make it ready for execution.
Compiler	A program which translates a high-level language program into computer's machine code or some other low-level language. Each high-level instruction is changed into several machine-code instructions. It produces an independent program which is capable of being executed by itself. This process is known as compilation.
Computer	A piece of hardware that can store and execute instructions (i.e. interpret them and cause some action to occur).
Configuration	A collection of items that bear a particular relation to each other (e.g. the data configuration of a system in which classes of data and their relationships are defined).
CORBA	Common Object Request Broker Architecture. An evolving framework being developed by the Object Management Group to provide common approach to systems interworking.
CPE	Customer Premises Equipment. The equipment—such as terminals and routers—found at customer sites. Put another way, not part of the public network.
CSMA/CD	Carrier Sense Multiple Access with Collision Detection—a method used in local area networks whereby a terminal station wishing to transmit *listens* and transmits only when the shared line is free. If two or more stations transmit at the same time, each backs off for a random time before re-transmission. Each station monitors its signal and if this is different from the one being transmitted, a collision is said to have occurred (collision detection). Each backs off and then tries again later.
Customer	A person or entity paying for a piece of software. There are many different types of customer and these are detailed in the main text.
Cyberspace	A term used to describe the world of computers and the society that gathers around them. First coined by William Gibson in his novel *Neuromancer*.
Data	Usually the same as information. Sometimes information is regarded as processed data.
Data compression	A method of reducing the amount of data to be transmitted by applying an algorithm to the basic data source. A decompression algorithm expands the data back to its original state at the other end of the link.
Data design	The design of the data structure needed by a particular software system.

Database	A collection of interrelated data stored together with controlled redundancy to support one or more applications. On a network, data files are organised so that users can access a pool of relevant information.
Database server	The machine that controls access to the database using client–server architecture. The server part of the program is responsible for updating records, ensuring that multiple access is available to authorised users, protecting the data and communicating with other servers holding relevant data.
DBMS	DataBase Management Systems—groups of software used to set up and maintain a database that will allow users to call up the records they require. In some cases, DBMS also offer report and application generating facilities.
DCE	Distributed Computing Environment. A set of definitions and components for distributed computing developed by the Open Software Foundation, an industry led consortia.
Deadlock	A condition where two or more processes are waiting for one of the others to do something. In the meantime, nothing happens. A condition (undesirable) that needs to be guarded against, especially with network protocols and in the design of databases.
Debugging	The detection, location and correction of bugs.
DECNet	Protocols and products from the Digital Equipment Corporation in wide use for computer interconnection. DECNet phase V is largely based on OSI protocols.
Design	(n) A plan for a technical artefact. A definition of the components that will be used to create an artefact along with details of their configuration, operation and deployment. (v) To create a design; to plan and structure a technical artefact.
Dimensioning	This is the process of assigning a given size to a given network connection. For instance, in Frame Relay, it is the step of finding the access link speed and committed information rate for a Private Virtual Circuit.
Distributed computing	In a move away from having large centralised computers such as minicomputer and mainframes, and bring processing power to the desktop. Often confused with distributed processing.
Distributed database	A database that allows users to gain access to records, as though they were held locally, through a database server on each of the machines holding part of the database. Every database server needs to be able to communicate with all the others as well as being accessible to multiple users.
Distributed processing	The distribution of information processing functions amongst several different locations in a distributed system.
DNS	Domain Name Service. The method used to convert Internet names to their corresponding Internet numbers.
DoD	The US Department of Defense. Notable for their sponsorship of the network that was to become the Internet.

Domain	Part of a naming hierarchy. A domain name consists of a sequence of names or other words separated by dots. Also, a part of a network.
Dongle	A security or copy protection device for commercial microprocessor programs. Programs query the dongle (a device that needs to inserted in the external port of the computer) before they will run.
E1	A standard transmission rate of 2.048 Mbps. Usually refers to a European leased line.
ECMA	European Computer Manufacturers Association—an association comprised of members from computer manufacturers in Europe, it produces its own standards and contributes to CCITT and ISO.
Encapsulation	A method whereby the data transfer capability of one protocol is used to transport the protocol information from another. This nesting allows two systems similarly configured to use the same protocol to communicate using a dissimilar network protocol.
E-mail	Common abbreviation for electronic mail. E-mail comes in many guises and has been popularised, to a large extent, through the growth of the Internet.
Engineering	A process of applying scientific and other information in specific ways to achieve technical, economic and human goals.
Enterprise Network	The collection of public and private switches, transmission links, subnetworks, management systems, network applications etc., that combine to provide an identifiable group of people with the communication service that they need to operate effectively.
Ethernet	A local area network (LAN) characterised by 10 Mbit/s transmission using CSMA/CD Carrier Sense Multiple Access with Collision Detection.
Fast packet switching	A new technology that differs from traditional packet switching. One differing aspect is that it transmits all data in a single packet format whether the information is video, voice or data.
Fault tolerance	A method of ensuring that a computer system or network is more resilient to faults or breakdowns, to avoid lost data and downtime. Differing applications achieve this and include processor duplication and redundant media systems.
FAQ	Frequently asked question. A set of files, available over the Internet, that provide a compendium of accumulated knowledge in a particular subject.
FDDI	Fibre Distributed Data Interface—an American National Standards Institute (ANSI) LAN standard. It is intended to carry data between computers at speeds up to 100 Mbit/s via fibre optic links. It uses a counter rotating Token Ring topology and is compatible with the first, physical, level of the ISO seven-layer model.
Feature interaction	Term used to describe knock-on effects in system design. Most commonly experienced when an 'enhancement' to one feature causes another to cease working (a feature that allows access to more databases causes the system configuration records to overwrite). In many ways, a fancy way of describing a design bug.

File server	A station in local area networks dedicated to providing file and data storage to other terminals in the network.
Flow Control	In a packet switched network, packets compete dynamically for the networks resources—storage, processing, transmission. Flow control is a mechanism for ensuring fairness and controlling congestion.
FTAM	File Transfer and Manipulation—a protocol entity forming part of the application layer enabling users to manage and access a distributed file system.
FTP	File Transfer Protocol. The Internet standard high-level protocol for transferring files from one computer to another. A widely used *de facto* standard.
FYI	For your information. These are Internet bulletins that answer common questions .
G.703	The CCITT standard for the physical and logical transmissions over digital circuit. Specifications include the US 1.544 Mbit/s as well as the European 2.048 Mbit/s that use the CCITT recommended physical and electrical data interface.
G.703 2.048	Transmission facilities running at 2.048 Mbits/s that use the CCITT Mbits/s recommended physical and electrical data interface.
Gateway	Hardware and software that connect incompatible networks, which enables data to be passed from one network to another. The gateway performs the necessary protocol conversions.
Gopher	One of a number of Internet based services that provide information search and retrieval facilities.
GOS	Grade of Service. The probability that a call or connection will be blocked. This relates to circuit switched networks, and is determined by the amount of switching and transmission equipment provided (per user) in the network.
GOSIP	Government Open Systems Interconnect Profiles. A UK initiative to help users procure open systems.
GUI	Graphics User Interface—an interface that enables the user to select a menu item by using a mouse to point to a graphic icon (small simple pictorial representation of a function such as a paint brush for shading diagrams etc.). This is an alternative to more traditional character-based interface where an alphanumeric keyboard is used to convey instructions.
Hardware	The physical equipment in a computer system. It is usually contrasted with software.
Hierarchical network	A network structure composed of layers. An example of this can be found in a telephone network. The lower layer is the local network followed by a trunk (long-distance) network up to the international exchange networks.
Host	Commonly used synonym for server.
Hostage data	Data which is generally useful but held by a system which makes external access to the data difficult or expensive.

HTML   Hyper Text Markup Language is used to describe formatting in web
                documents. It is defined in the Internet's RFC series.

HTTP   Hyper Text Transfer Protocol. This is the basic protocol that underlies
                the World-wide Web. It is a simple, stateless request-response protocol,
                defined in the Internet's RFC series.

IAB   Internet Architecture Board. The influential panel that guides the
                technical standards adopted over the Internet. Responsible for the
                widely accepted TCP/IP family of protocols. More recently, the IAB
                have accepted SNMP as their approved network management protocol.

IEEE   The Institute of Electrical and Electronic Engineers. US based profes-
                sional body covering network and computing engineering.

IEE   UK equivalent of the IEEE.

IEEE 802.3   The IEEE's specification for a physical cabling standard for LANs as
                well as the method of transmitting data and controlling access to the
                cable. It uses the CSMA/CD access method on a bus topology LAN
                and is operationally similar to Ethernet. Also see OSI.

IEEE 802.4   A physical layer standard that uses the Token Ring passing access
                method on a bus topology LAN.

IEEE 802.5   A LAN physical layer standard that uses the Token Ring passing
                access method on a ring topology LAN. Used by IBM on its Token
                Ring systems. Also see OSI.

Implementation The process of converting the notation used to express detailed design
                into specific hardware and software configurations. Implementation
                also denotes the task of bringing together the various systems compo-
                nents to get the system working (also known as commissioning).

Information   A computer based processor for data storage and/or manipulation
processor   services for the end user.

Information   Any method or procedure which is used for the recovering of
retrieval   information or data which has been stored in an electronic medium.

Information   A rather loosely defined term that is usually taken to cover the
technology   technology relevant to providing information services for an or-
                ganisations. Generally a mix of software, hardware, office systems
                databases, networks and computing.

Intelligent   A terminal which contains a processor and memory with some level
terminal   of programming facility. The opposite is a dumb terminal.

Interface   The boundary between two things, typically two programs, two
                pieces of hardware, a computer and its user, a project manager and
                the customer.

Internetworking Specialised form of interworking where the interaction involves two
                or more networks.

Interworking   Generic interaction between two entities to achieve operational,
                syntactic and semantic integrity of information.

ISDN	Integrated Services Digital Network—an emerging end-to-end CCITT standards for voice, data and image services. The intention is for ISDN to provide simultaneous handling of digitised voice and data traffic on the same links and the same exchanges.
ISO	Commonly believed to stand for International Standards Organisation. In fact ISO is not an abbreviation—it is intended to signify commonality (from Greek *Iso* = same) the ISO is responsible for many data communications standards. A well-known standard produced by ISO is the seven-layer Open Systems Interconnection (OSI) model.
ISO 9001	The standard most often adopted to signify software quality. In reality it assures no more than a basic level of process control in software development. The focus on process control (documentation, procedures etc.) sometimes promotes bureaucracy in ISO 9001 accredited organisations.
Isochronous	Data transmission in which a transmitter uses a synchronous clock but the receiver does not. The received detects messages by start/stop bits as in asynchronous transmission.
ISP	Internet Service Provider. A supplier to the general public of points of presence on the Internet.
Java	A programming language and environment for developing mobile code applicatons. Java is a subject of the C++ language and is widely used to provide mobile code application for use over the Internet.
JFDI	Just do it. An approach to managing software projects that spurns too much introspection in favour of action.
Key	The record identifier used in many information retrieval systems (i.e. database keys).
LAN	Local area network—a data communications network used to interconnect data terminal equipment distributed over a limited area.
Language	An agreed-upon set of symbols, rules for combining them and meanings attached to the symbols that is used to express something (e.g. the Pascal programming language, job-control language for an operating system and a graphical language for building models of a proposed network).
LAN Manager	A network operation system developed by Microsoft for PCs running IBM's OS/2.
Legacy system	A system which has been developed to satisfy a specific requirement and is, usually, difficult to substantially re-configure without major re-engineering.
Lifecycle	A defined set of stages through which a piece of software passes over time—from requirements analysis to maintenance. Common examples are the waterfall (for sequential, staged developments) and the spiral (for iterative, incremental developments). Lifecycles do not map to reality too closely but do provide some basis for measurement and hence control.

MAC	Media Access Control. This controls access to the shared transmission medium by framing/deframing data units, error checking and providing access rights. Conceptually, the MAC is part of data link control and is important in the operation of LANs.
Maintenance	Changes to a network after its initial development; also called evolution. In practice, it is the task of modifying (locating problems, correcting or updating etc.) hardware configuration and (more often) software systems after they have been put into operation.
Method	A way of doing something. It is generally a defined approach to achieving the various phases of the lifecycle. Methods are usually regarded as functionally similar to tools (e.g. a specific tool will support a particular method).
MHS	Message Handling Service—the protocol forming part of the applications layer and providing a generalised facility for exchanging messages between systems.
MIB	Management Information Base. The data schema that defines information available in an SNMP enabled device. MIB (now at version 2) is defined in RFC 1213.
Mips	Millions of instructions per second—one measure of a computer's processing power is how many instructions per second it can handle.
Model	An abstraction of reality that bears enough resemblance to the object of the model that we can answer some questions about the object by consulting the model.
Modelling	Simulation of a system by manipulating a number of interactive variables; can answer 'what if …?' questions to predict the behaviour of the modelled system. A model of a system or subsystem is often called a prototype.
Modem	MOdulator–DEModulator, Data Communications Equipment that performs necessary signal conversions to and from terminals to permit transmission of source data over telephone and/or data networks.
Modularisation	The splitting up of a software system into a number of sections (modules) to ease design, coding etc. Only works if the interfaces between the modules are clearly and accurately specified.
MPLS	Multi-protocol label switching. A means of identifying packets in a data stream.
Multiplexing	The sharing of common transmission medium for the simultaneous transmission of a number of independent information signals, see Frequency Division Multiplexing (FDM) and Time Division Multiplexing (TDM).
Named Pipes	Part of Microsoft's LAN Manager—an interface for interprocessing communications and distributed applications. An alternative to NetBios designed to extend the interprocess interfaces of OS/2 across a network.
NetBios	Network Basic Input/Output System—an IBM developed protocol. It enables IBM PCs to interface and have access to a network.
NetWare	A Novell Inc LAN operating system and associated products. Novell are a major player in the world LAN server market.

Network	A general term used to describe the inter-connection of computers and their peripheral devices by communications channels. For example Public Switched Telephone Network (PSTN), Packet Switched Data Network (PSDN), local area network (LAN), wide area network (WAN).
Network interface	The circuitry that connects a node to the network, usually in the form of a card fitted into one of the expansion slots in the back of the machine. It works with the network software and operating system to transmit and receive messages on the network.
Network management	A general term embracing all the functions and processes involved in managing a network, and include configuration, fault diagnosis and correction. It also concerns itself with statistics gathering on network usage.
Network topology	The geometry of the network relating to the way the nodes are interconnected.
NFS	Network file system. A method, developed by Sun microsystems, that allows computers to share files across a network as if they were local.
Non-proprietary	Software and hardware that is not bound to one manufacturer's platform. Equipment that is designed to the specification that can accommodate other companies' products. The advantage of non-proprietary equipment is that a user has more freedom of choice and a larger scope. The disadvantage is when it does not work, you may be on your own.
NIC	The Network Information Centre. Source of much information on the Internet and related networking issues.
NT	Network Termination. A piece of equipment (usually on a users' premises) that provides network access via a standard interface. Also, an established Microsoft operating system (with NT = new technology).
Object	An abstract, encapsulated entity which provides a well-defined service via a well-defined interface.
Object oriented	A philosophy that breaks a problem into a number of co-operating objects. Each object has defined properties (e.g. it can inherit features from another object). Object oriented design is becoming increasingly popular (e.g. in the specification of network management systems).
Object program	The translated versions of a program that has been assembled or compiled.
ODP	Open Distributed Processing. One of a number of organisations (most of which have the word 'open' in their title) which provide standards and/or components that allow computers from different vendors to interwork.
OMG	Object Management Group. The OMG is responsible for the CORBA initiative.
OMNI	Open Management Inter-operability. An ISO based network management standards body. Responsible for OMNIPoint, which includes the CMIS and CMIP standards for connection between network elements and network management systems.

Operating system	Software such as MVS OS/2 (IBM), VMS (DEC), DOS, Unix that manages the computer's hardware and software. Unless it intentionally hands over to another program, an operating system runs programs and controls peripherals.
OSCA	Open Systems Cabling Architecture—structured cabling system, primarily for local networks.
OSF	Open Systems Foundation. An organisation that provides generic Unix based software components.
OSI	Open Systems Interconnection—the ISO Reference Model consisting of seven protocol layers. These are the application, presentation, session, transport, network, link and physical layers. The concept of the protocols is to provide manufacturers and suppliers of communications equipment with a standard that will provide reliable communications across a broad range of equipment types.
OSPF	Open Shortest Path First. A routing protocol for TCP/IP networks.
Packet switching	The mode of operation in a data communications network whereby messages to be transmitted are first transformed into a number of smaller self-contained message units known as packets. Packets are stored at intermediate network nodes (packet-switched exchanges) and are reassembled unto a complete message at the destination. A CCITT recommendation standard for packet switching is X.25.
PAD	Packet Assembler and Disassembler—a device used in the X.25 packet switched network, permitting terminals which cannot interface with the network to do so. PAD converts terminals data to/from packets and handles call set up and addressing.
Parameter	A variable whose value may change the operation but not the structure of some activity (e.g. an important parameter in the productivity of a program is the language used).
Pathological	Use of a data set that is at (and beyond) the extremes of the typical. Pathological data is frequently used to test a system's response to exceptional conditions.
Peer to peer	Communications between two devices on an equal footing, as opposed to host/terminal, or master/slave. In peer to peer communications both machines have and use processing power.
Pipe	Installed in most operating systems, a pipe is a method used by distributed programs to communicate with each other. One of OS/2's attributes is that pipes can be created quickly and easily. When a program, sends data to a pipe, it is transmitted directly to the other program without ever being written onto a file.
Pixel	Picture element—the smallest discrete element making up a visual display image.
Point-to-point	Direct link between two points in a network or communications link.
Polling	Process of interrogating terminals in a multipoint network in turn in a prearranged sequence by controlling the computer to determine

whether the terminals are ready to transmit or receive. if so, the polling sequence is temporarily interrupted while the terminal transmits or receives.

Port
A device which acts as an input/output connection. Serial port or parallel port are examples.

Procedure
A method or set of steps defining an activity; technically, a program that can be executed as a subactivity by another program.

Process
Technically, a procedure that is being executed on a specific set of data; more generally, a procedure for doing something that is actually being carried out.

Program
A set of instructions for a computer, arranged so that when executed they will cause some desired effect (such as the calculation of a quantity or the retrieval of a piece of data).

Programming language
An artificial language constructed in such a way that people and programmable machines can communicate with each other in a precise and intelligible way. Fortran, Cobol and C are three languages that account for most of the deployed software systems at present.

Proprietary
Any item of technology that is designed to work with only one manufacturer's equipment. The opposite of the principle behind Open Systems Interconnection (OSI).

Protocol
A set of rules and procedures that are used to formulate standards for information transfer between devices.

Prototype
A scaled-down version of something, built before the complete item is built, in order to assess the feasibility or utility of the full version.

PSTN
Public Switched Telephone Network—the public telephone system providing local, long-distance and international telephone services. Widely used (with modems) for many other data services.

PTT
Postal, Telegraph and Telephone—the administrative authority in a country that controls all postal and public telecommunication services in that country. Same as PNO—Public Network Operator.

Quality assessment
A systematic and independent examination to determine whether quality activities and related results comply with planned arrangements and whether these arrangements are implemented effectively and are suitable to achieve objectives.

Quality surveillance
The continuing monitoring and verification of the status of procedures, methods, conditions, processes, products and services and the analysis of records in relation to stated references to ensure that specified requirements for quality are being met.

Quality system
The organisational structure, responsibilities, procedures, processes and resources for implementing quality management.

Queuing
When a frame or packet is to be transmitted on a link, it may have to wait because another frame is being processed in front of it. The frame is placed in a buffer until the transmitter is free. Hence queuing systems (i.e. packet switched systems) require buffers (matched to

	load and capacity) and introduce delay (as opposed to circuit switching systems, which block).
Requirements analysis	The analysis of a user's needs and the conversion of these into a statement of requirements, prior to specification.
Resolve	Translate an Internet name into its equivalent IP address or other DNS information.
RFC	Request for comment. A long-established series of Internet informal documents widely followed by commercial software developers. RFCs tend to provide the implementation detail to supplement the more general guidance of ISO and other formal standards. The main vehicle for the publication of Internet standards, such as SNMP.
RIP	Routing Information Protocol, a standard gateway protocol defined in RFC 1388 that uses message broadcasts to a destination based on hop count.
RMON	Remote Monitoring management information database. Developed by the IETF, this extension of SNMP MIB 2 provides a means for tracking, storing and analysing remote network management information.
ROSE	Remote Operations Service Element a protocol code forming part of the applications layer, providing a facility for initiating and controlling operations remotely.
Routers	A router operates at level 3 of the OSI model. Routers are protocol specific and act on routing information carried out by the communications protocol in the network later. A router is able to use the information it has obtained about the network topology and can choose the best route for packet to follow. Routers are independent of the physical level (layer 1) and can be used to link a number of different network types together.
Routing	The selection of a communications path for the transmission of information from source to destination.
SASE	Specific Application Service Element—a collection of protocol codes forming part of the application layer for specific services such as file and job transfers.
SCSI	Small Computer System Interface. A bus-independent standard for system level interfacing between a computer and an intelligent device, e.g. an external disk (pronounced—scuzzy).
SDH	Synchronous Digital Hierarchy. See Sonet.
Service	An independently useful, and well-defined function.
Server	An object which is participating in an interaction with another object, and is taking the role of providing the required service.
Session	The connection of two nodes on a network for the exchange of data—any live link between any two data devices.

Signalling	The passing of information and instructions from one point to another for the setting up or supervision of a telephone call or message transmission.
SNA	Systems Network Architecture—an IBM layered communications protocol for sending data between IBM hardware and software.
SNMP	Simple Network Management Protocol—consists of three parts: Structure of Management Information (SMI), Management Information Base (MIB) and the protocol itself. The SMI and MIB define and store the set of managed entities, SNMP transports information to and from these entities.
SMTP	Simple mail transfer protocol. The Internet standard for the transfer of mail messages from one processor to another. The protocol details the format and control of messages.
Software	All programs (plus documentation) which are associated with a computer or computer based system, as opposed to hardware which is the physical equipment. It also refers to programs, data, designs for programs, specifications and any of the other information that is relevant to a particular set of executable computer instructions (either existing or planned).
Software engineering	The development and use of systematic strategies (themselves often software based) for the production of good quality software within budgets and to timescales.
Sonet	A synchronous optical transmission protocol (UK equivalent is SDH). Sonet is intended to add and drop lower bit rate signals from the higher bit rate signal without needing de-multiplexing. The standard defines a set of transmission rates, signals and interfaces for fibre optic transmission.
Specification	A description of a system or program that states what should be provided but does not necessarily provide information on exactly how the system or program will work.
SQL	Structured Query Language. Probably the most widely used database access language.
Store and forward	A technique used in data communications in which messages packets are stored at an intermediate node in a network and then forwarded to the next routing point where an appropriate line becomes free.
Switched Network	A network which is shared by several users, any of whom can establish communication with any other by means of suitable inter-connections (switching) operations.
Switching	Process by which transmissions between terminals are interconnected, effected at exchange at nodal points exchanges in the network.
Synchronisation	The actions of maintaining the correct timing sequences for the operation of a system.
Synchronous transmission	Transmission between terminals where data is normally transmitted in blocks of binary digit streams and transmitter and receiver clocks are maintained in synchronism.

Syntax	The set of rules for combining the elements of a language (e.g. words) into permitted constructions (e.g. phrases and sentences). The set of rules does not define meaning, nor does it depend on the use made of the final construction.
System	A collection of independently useful objects which happen to have been developed at the same time. Also, a collection of elements that work together, forming a coherent whole (e.g. a computer system consisting of processors, printers, disks etc.).
System design	The process of establishing the overall architecture of a software system.
Systems program	One of the programs which control the performance of a computer system (e.g. computer or monitor).
Systems software	Software, such as an operating system, concerned mainly with 'house-keeping' tasks, managing the hardware resources, etc. It is usually contrasted with applications software.
T1	A standard transmission rate of 1.554 Mbps. Usually refers to a US leased line.
TCP/IP	Transmission Control Protocol/Internet Protocol—a set of transport internetworking protocols that have become a *de facto* networking standard. Commonly used over X.25 and Ethernet wiring, they are viewed as two of the few protocols available that are able to offer a true migration path towards OSI. TCP/IP operates at the third and fourth layers of the OSI model (network and transport respectively).
Telecommunications	The general name given to the means of communication information over a distance by electrical and electromagnetic methods. The transmission and reception of information by any kind of electromagnetic system.
Telnet	An TCP/IP based application that allows connection to a remote computer.
Throughput	A way of measuring the speed at which a system, computer or link can accept, handle and output information.
Topology	A description of the shape of a network, for example star, bus and ring. It can also be a template or pattern for the possible logical connections onto a network.
TP	Transaction Processing. Concerned with controlling the rate of enquiries to a database. Specialist software—known as a TP monitor—allows a potential bottleneck to be managed.
Transmission	The act of transmitting a signal by electrical/electromagnetic means over a communications channel.
UMTS	Universal Mobile Telephony Service. This is the concept that users will be able to access all of their network services irrespective of location. The goal of UMTS is to provide a network based around the individual – the individual registers their access device with the network rather than having to find a fixed access device.
UPT	Universal Personal Telephony.

UUCP	Unix–Unix Communication Protocol. A basic mechanism that allows computers running the Unix operating system to inter-operate.
Validation	The process of checking a specific piece of lifecycle notation, and the conversion from one piece of notation to another.
Vendor independent	Hardware or software that will work with hardware and software manufactured by different vendors—the approximate opposite of proprietary.
Virtual circuit	A logical connection across a network (e.g. the transmission path through an X.25 packet-switched data network established by exchange of set up messages between source and destination terminals).
Virtual device	A module allocated to an application by an operating system or network operating system, instead of a real or physical one. The user can then use a computer facility (keyboard, memory, disk or port) as though it was really present. In fact, only the operating system has access to the real device.
Virus	A program, passed over networks, that has potentially destructive effects once it arrives. Packages such as Virus Guard are in common use to prevent infection from hostile visitors.
VPN	Virtual Private Network. A combination of public and private resources that has been combined to give the user a network that looks like a coherent resource, suited to their particular needs. To all intents and purposes, a VPN is an enterprise network.
Waterfall	The name for the 'classical' system development lifecycle, so named because the chart used to portray it suggested a waterfall.
Window	A flow control mechanism the size of which determines the number of data units that can be sent before an acknowledgement of receipt is needed,and before more can be transmitted.
Windows	A way of displaying information on a screen so that users can do the equivalent of looking at several pieces of paper at once. Each window can be manipulated for closer examination or amendment. This technique allows the user to look at two files at once or even to run more than one program simultaneously.
World Wide Web	An Internet based project that has provided an intuitive and powerful means of navigating a very large information space. The Web has become almost synonymous with the client packages (browsers) such as Netscape and Internet Explorer that allow a user to traverse the Internet via hypertext links to access multimedia information.
Workstation	A networked personal computing device with more power than a typical PC. Usually a Unix machine capable of running several tasks at the same time.
Worm	A computer program which replicates itself—a form of virus. The Internet worm was probably the most famous—it successfully, and accidentally, duplicated itself across the entire system.
X.25	A CCITT standard for packet format for public data networks. An X.25 network can carry other types of data (e.g. IP, DECNet, XNS). Most public operators offer an X.25 service.

X.400   A store and forward Message Handling System (MHS) standard that allows for the electronic exchange of text as well as other electronic data such as graphics and fax. It enables suppliers to interwork between different electronic mail systems. X.400 has several protocols, defined to allow the reliable transfer of information between User Agents and Message Transfer Agents.

X.500   A directory services standard that permits applications such as electronic mail to access information, which can either be central or distributed.

XML   eXtensible Markup Language. Developed by the Worldwide Web Consortium for describing and exchanging information between applications and environments that are based on different technologies and have different semantic representations. It uses the Internet HTTP (HyperText Transport Protocol) to say how information should be presented and serves to identify the content.

XNS   Xerox Network Systems. A protocol designed for interconnecting PCs. Used (in several guises) by Novell, Ungermann-Bass, Banyan and other LAN suppliers.

X/Open   An industry standards consortium that develops detailed system specifications drawing on available standards. X/Open owns the Unix trademark and thereby brings focus to its various flavours (e.g. HP-UX, AIX from IBM, Solaris from SUN etc.).

Index

Printed and bound by CPI Group (UK) Ltd, Croydon, CR0 4YY

27/10/2024

14580285-0004